BURNING THE PAGE

THE EBOOK REVOLUTION AND THE FUTURE OF READING

JASON MERKOSKI

Published by Sourcebooks, Inc.
P.O. Box 4410, Naperville, Illinois 60567-4410
(630) 961-3900
Fax: (630) 961-2168
www.sourcebooks.com

Library of Congress Cataloging-in-Publication Data

Merkoski, Jason.
Burning the page : the ebook revolution and the future of reading / Jason Merkoski.
pages cm
Includes indexes.
(pbk. : alk. paper) 1. Electronic books. 2. Books and reading. 3. Books and reading—Forecasting. I. Title.
Z1033.E43M47 2013
070.5'73—dc23

2013017012

Printed and bound in the United States of America.
VP 10 9 8 7 6 5 4 3 2 1

Disclaimer

This book is based on my recollections of the time I spent working on the front lines of the ebook revolution. It also contains my opinions and predictions about the development of ebooks: where we have been and where we are heading. The content of this book is based on my opinions and personal experiences and on publicly reported data about Amazon and other companies.

The opinions, predictions, and content expressed in this book and in the online portions of this book via links at the end of each chapter are entirely my own and are not those of Amazon, its affiliates or subsidiaries, or Sourcebooks, Inc. Neither is responsible for the contents of this book or the online portions of this book.

"This is a book, my darling. A magic book..."
Her eyes got wide. The book began to talk to her.

—*Neal Stephenson, The Diamond Age*

Contents

How It Started

This is a story about ebooks. It's a story about Google, Jeff Bezos, and the ghost of Gutenberg. It's the true story of the ebook revolution—what ebooks are and what they mean for you and me, for our future, and for reading itself.

I was fortunate to find myself in the midst of this revolution from the beginning. As one of the founding members of Amazon's Kindle team, I was part of a small group of people who started the revolution in reading, who set out to change the way the world reads. I got in on Kindle on the ground floor with the goal to make all books in all languages downloadable in less than sixty seconds. At Amazon, I was an engineering manager, program manager, product manager, and evangelist. This gave me a big-picture perspective on ebooks, on how they were created, sold, and read.

I not only learned all about ebooks during this time, but also invented many of the features we now take for granted. If you use a Kindle, I had a hand in shaping it. During the half decade I was there, Kindle succeeded beyond our imaginations. It succeeded in sales, in popularity, and in changing how we read. We wanted to change the world, and we did.

We started an ebook revolution.

When I talk about revolution, I'm not referring to political or regime-changing revolutions like the Khmer Rouge reign of terror or the French Revolution. I'm not talking about massacres and beheadings.

I'm talking about movements that change how we live, think, and perceive the world around us, such as the Industrial Revolution or the civil rights movement. I'm talking about a technical revolution, a scientific revolution, a social revolution.

Revolution is what you get when technology and culture collide.

The ebook revolution is changing all the rules for reading and writing. It's changing entertainment, and it's allowing our culture to immortalize itself through digitization. Ebooks can do things print books never could. You can now download an ebook as fast as you can call a friend on the phone. You can fit a library in your pocket. You can send a thousand ebooks to a village school in Africa, free of worries about quarantines, customs, bribes, and tangled parachute lines. You can read an ebook at the same time as me, halfway around the world, and we can discuss it together and share our comments and thoughts on it.

Books are what make us human, what set us apart from all other animals. And by connecting with books, by crossing the chasms of culture and language through them, humanity itself becomes connected. Reading—once a primarily solitary and individual activity—can now be social on a planetary scale.

Ebooks have the power to ignite us.

These are interesting times for reading. In the 1960s, the future was "plastics," as anyone who's seen *The Graduate* knows, but these days, the future is digital. The future will have meshed and interconnected devices like e-readers that respond to you wherever you are. In some ways, the future is already here.

For example, you can start reading on an e-reader and later continue reading where you left off on your phone. The ebook doesn't care what device you're reading it on; it just seamlessly integrates with you wherever you are. And you can use ebooks as a canary in the coal mine to see where the future is going—not just for digital content, but for our digital lives, as well.

Are digital books the death knell for printed books, or will they breathe new life into them? Will ebooks give you features that enhance the reading experience or distract you from it? Will these experiments about the form of the book kill it? Or will they elevate books to a new place of honor in our culture? How will we change intellectually and emotionally as our reading habits change?

Tough questions.

Though I'm an ebook inventor and technologist, I'm also a humanist. Ebooks will never quite smell as nice as musty library volumes or books from your childhood that still have forgotten lilacs pressed between the pages from so many summers ago. At best, e-readers will smell like formaldehyde and plastic or the metal tang of an overheated battery.

If you're like me, you're passionate about books as things you can touch, that you can dog-ear or annotate, and that have covers you've come to enjoy. You and I both worry about what it means to put our personal libraries onto one gadget and then what would happen if we dropped it in the bathtub or stepped on it or put it with the laundry inside the washing machine. If you are like me, you have more books than you have friends, no matter what Facebook tells you about your social network.

That said, although I love print books with all my heart, I also believe in the power of ebooks. I spent five long years at Amazon inventing ebook technology, launching devices, and creating crazy new ways of reading. Because I was on the team for so long, I became the closest there was to an ebook shaman, a tribal elder who could talk to all the people who joined Amazon after me about the early days of Kindle, provide the inside scoop. So I'm going to give you the same inside scoop—but about ebooks as a whole, not just the Kindle. This book will explain how ebooks came to be, and once you know that, you can look ahead into the future of reading, communication, and human culture.

After all, sometimes the best way to see where you're going is first to look back at where you came from.

So where did I come from? If there is a story to my life, it's a story of books.

Some people keep a stack of magazines in their bathroom, but I keep a stack of e-readers, like a Sony Reader and a Nook and an iPad. All different kinds. I also keep a stack of books near my bed—twenty or so heaped high on top of one another, half opened. I have books to read wherever I go, even audiobooks when I'm driving. I have more than 4,000 printed books, and I can't even count how many ebooks I own. Fiction, nonfiction: I love it all.

I was born in New Jersey, midway between the Garden State's blueberry fields and Atlantic City's casinos. My grandfather never learned how to read. He was a truck driver in New Jersey who barely managed to scrape up enough quarters over the years to send my father to college. My father worked at a newspaper and always came home smelling of newsprint and the latest headlines.

What can I say? Ink runs in my blood. I was a shy guy at school, so I would often read—before school, during school, after school—and in retrospect, it seems like most of the time I spent at school was inside a book.

I went to the Massachusetts Institute of Technology (MIT) and initially trained as a physicist because I wanted to know how the universe works. But I realized that math is more universal, so I changed majors. Then I realized that all I was doing with math was using symbols, like grammar. Math is a language, but you can't tell stories with it. The English language is more expressive. So I started writing instead.

After graduating, I wrote on nights and weekends for ten years, working on a colossal sprawling novel set in the 1930s. I had a bunch of day jobs during this time—I ran technology for a number of companies on the East Coast, and I built Motorola's first e-commerce system. But during the dot-com boom, I took a sabbatical from work and moved to New Mexico to finish writing my novel. Everyone was making their dot-com fortunes, but I was writing about the Great Depression!

When it was done, the book was a million words long and illustrated. I put it online as the web's first internet novel; it was an ebook before there were ebooks. You could turn pages in your browser, annotate words or sentences you liked, and bookmark pages. If you stopped and later wanted to resume reading, you'd continue where you left off. I created all these features from scratch, little knowing that I was laying the groundwork for the first e-readers.

Halfway through the first decade of the twenty-first century, I was living in the wilds of New Mexico again, and I heard that Amazon and Google were working on book digitization projects. As a word lover and text aficionado, I was intrigued and applied for jobs at both companies. I went through grueling interviews with each, basically locked in a conference room all day. You spend an hour with each person who comes in to interview you, and you write code on a whiteboard or draw architecture diagrams.

It's tough and rigorous: people sometimes leave the interview crying, knowing they've failed, and are escorted outside by security. Not only are the interviews hard in general, but many tech companies also throw in "bar-raisers" who ask you questions so hard that you're supposed to feel like you've failed the interview.

There I was, dressed in cowboy boots and a trippy paisley shirt, talking to overworked engineers with barbecue stains on their T-shirts. I talked about linguistics, about Sanskrit, a language I had been teaching myself; I talked about publishing, my love of books, and writing. Finally, I talked about my technical expertise and my visions for the future. I aced my interviews and played Amazon against Google during salary negotiations.

Amazon had a director call to sell me on working there. I still remember where I was the day I spoke to him—sitting on my floor in a patch of sunlight, listening to his voice crackle from more than a thousand miles away on my rural phone line. I lived in a remote place, at the very end of the power and phone grids. Through the static, he hinted that Amazon was working on a secret ebook project, and he offered me any position I wanted on it.

I chose the hardest one—working on a team that had to invent a way of making ebooks from physical books. I was in Seattle two weeks later at new employee orientation, watching an overhead projection of Jeff Bezos's head welcoming me to work, telling me to have fun and make history. I joined the Kindle team, and for a few years, I worked in a modern version of Gutenberg's workshop. Most of what we did on Kindle was digital, so you can't physically see it; it's like the part of an iceberg below the ocean. The outward form of what you see is a piece of plastic and metal, a Kindle.

What you don't see are the warrens of cubicles from which Amazon reps called publishers every day asking for more books. You don't see all the engineers, all the code they've written to make payments to book publishers happen every month or to manage wireless downloads or to audit the books in users' libraries to make sure they're still there. The Kindle itself is just the tip of the iceberg, and its true workings are invisible. That's exactly how Amazon wanted it to be.

Yes, I did have fun at Amazon, and I made history. I first joined a team that built the electronic books for Kindle, but I went on from there to do it all. I invented some of the technology used in ebooks and launched the first few Kindles. I traveled to book fairs in New York and London and Frankfurt to evangelize ebooks. I watched ebooks being made in the Philippines and supervised the assembly of Kindles in China. I talked to the White House, former presidents, and astronauts about ebooks. I worked with *Wired* magazine and Random House. I talked about the future of books with the commissioner of the NFL and the founder of Wikipedia.

I also joined a crazy tribe whose members share their word-sparks with one another. And by this I don't mean just Amazon: members of this tribe come from all avenues of publishing. A special kind of person is drawn to publishing—they're often idealistic, as their decisions are made for the written word itself, for something greater than themselves, some spark or idea they want to share with others. These kinds of people are the innovators, the idealists, and they are a vital part of this book because they're revitalizing how we read. They're breathing new life into books.

This book talks about publishing, authorship, and the myriad ways we engage with the written word. While my perspective is rooted in my experiences working at Amazon, this book also discusses Apple and Google and publishers large and small. Like others at Amazon, I took my turn spinning the Kindle flywheel, from when it was just a glint of an idea in Jeff Bezos's eye to the whirlwind it is now, a phenomenon fanning the flames of reading around the globe. This book is the story not just of Kindle, but of the ebook revolution itself—what it is, where it's headed, and what it means for all of us, for good and for bad.

Introduction to "Bookmarks"

In the ebook revolution, the nature of books and the reading process is changing. Ebooks include everything useful that came before them in books. They add to the reading experience, not detract from it. But many now-familiar elements of print books are going the way of the dodo. Some will die out entirely; others will morph into something new. In part this is good—who's going to miss paper cuts, after all? But with this change also comes the loss of familiar friends.

The "bookmark" at the end of each chapter takes a look at an element of print books we have come to love or loathe and how it will be affected, transformed, or eliminated by the move to ebooks. As used here, the term "bookmark" is kind of a visual pun. Not only does it refer to an artifact from traditional print books, but each "bookmark" also is a small interlude that describes the ways books have indelibly marked our lives and our culture of reading. These sections are at once sentimental and speculative and appear throughout the book like bookmarks between chapters.

I will explain later in this book that I think there's really just one book, the book of all human culture. I'll describe what this one book might look like—as a sort of Facebook for Books, where all books can interact and link to one another in the same way that we ourselves are linked together on Facebook, as friends, coworkers, and family. We don't have this singularly hyperlinked book yet, but in an effort to build it, I'll invite you to talk with me and other readers throughout this book.

At the end of each "bookmark," you'll see a link you can—and should!—click to continue the conversation online. And I do encourage you to click each link; it lets you into a social reading app that connects you (via Facebook or Twitter) with other readers, with me, and with surprises all along the way. This app is simple to install and unlocks the brave new world of what I call "Reading 2.0." It's a world that combines a conversation with the author, a virtual book club, and a thoughtful friend who brings you special notes and treats. Please note: the web page is an independent site maintained by me and is independent of this book's publisher.

Clicking the link at the end of each "bookmark" unlocks a sequence of surprises and gifts—starting with a personalized autograph, bonus chapters, unexpected objects falling out from between the "pages" of the book, ways to carry on the conversation with other readers, and a personalized message upon completing the book. You need to click each link to unlock all the surprises.

I look forward to talking to you, because the greatest revolutionaries in the ebook revolution are the readers. We're all part of this revolution, and we're all mourning the culture of print books in our own ways. Everyone I talk to cares about the written word and has a strong opinion on where books are headed. Everyone has a story about books and ebooks and how they've changed our lives. So what's your story? Click the link below to share it with me and other readers—and get your autograph too!

http://jasonmerkoski.com/eb/1.html

The History of Books

If you've never used a Kindle, imagine a device the size of a book that can wirelessly download ebooks from Amazon's online store. These ebooks can be read just like regular books—you can turn pages, add bookmarks, see the cover, and go to the table of contents. But unlike regular books, ebooks also let you resize the text to make it bigger or smaller. Ebooks let you look up a word right there on the page to see its definition. Kindles are part computer, part book, and part cloud.

The packaging on the original Kindle box shows an illustrated history of the written word. Starting from the left-hand side, you see symbols in hieroglyphics and cuneiform, then you see Greek and Roman letters carved in stone, then woodblock medieval printing, finally followed on the right-hand side of the box by letters in modern alphabets on paper. The story of the written word is a story of evolution. In fact, the history of printing is one of a decline in durability and a rise in convenience.

Printing started 6,000 years ago with cuneiform tablets from the Middle East. These were created from wedges carefully cut from mud, fired in a kiln, and made into tablets. The process had more in common with sculpture than writing, but it was durable. We're still uncovering clay tablets from all over that region. I'd like to think that printing was invented to tell the stories of noble heroes and the elder goat-gods with their white beards, but no, most of these tablets were just bills and invoices. For example, in October 2012, an archive of 24,000 cuneiform business documents was found in central Turkey. It's a 6,000-year-old hoard of checks, tax forms, and loan notes.

Almost as old as clay tablets is papyrus, which is made from woven reeds like those that grow along the Nile. They're not as durable as clay and gradually decay, although in the Egyptian desert, you can still find fragments of papyrus preserved by desert sands and dust through the subsequent millennia. As writing boomed, the supply of reeds started dwindling, and in the fifth century BC, a new type of writing technology was developed, in which animal skins were made into parchment scrolls. Parchment lasts about a thousand years, and being made of animal skin, it's quicker to decay than papyrus and quick to crack, as anyone who owns a leather jacket or suede pants knows.

Paper was invented next, an even more convenient technology for printing, because wood pulp could be mashed up, laid out on racks to dry, and then cut into many thin sheets. It was much cheaper to make than any previous technology, but less durable. Even now, almost 2,000 years after its invention, paper only lasts about 500 years at best before yellowing and brittling to dust. Even the use of metal salts to make more durable acid-free paper isn't a recipe for immortality.

Over the last millennium, there have been other regional innovations in print technology, such as the use of palm leaves in Southeast Asia or birchbark among some of the Native Americans, but by and large, human civilization has predominantly used paper until now.

The history of book printing is wormy with false starts. For example, woodblock printing emerged around 200 AD in China before being rediscovered in Europe more than a thousand years later. Likewise, movable type was discovered and used to print books in Korea seventy-five years before it was rediscovered by Johannes Gutenberg, the credited inventor of modern printing. But it wasn't a single invention alone that sparked the blossoming of books in the Middle Ages. Gutenberg combined many inventions including moveable type, as well as the printing press and oil-based inks. The combination of all of these allowed book printing as we know it to succeed.

We don't know how he came up with these ideas and merged them together. In fact, we don't even know what Gutenberg looked like. The earliest illustrations of him didn't emerge until well after he died.

Except for his innovation, he's an almost absolute enigma, except for the occasional lawsuit filed against him. It would have been amazing if Gutenberg or one of his workers had written about making the first books, but they never did, or if so, their writing didn't endure. There's no written record of Gutenberg's workshop, but I imagine it would have been a lot like where newspapers were once printed, before the linotype and lead type were replaced by photography and digital typesetting.

The technology Gutenberg used in the 1450s was almost the same as the newspaper technology at my father's company three hundred years later in the latter half of the twentieth century. When I was a kid, I used to visit my father's newspaper on weekends. I would see enormous linotype machines that looked like a cross between typewriters and church organs, overheated machines belching steam while their operators sat with their burly 1970s mustaches and sweat-stained T-shirts, working to produce metal type.

The type would then be put into racks and ratcheted in with wrenches. Each line of the newspaper would be set with spacers between lines, and then the whole rack would be moved on an enormous system of pulleys to a room where it would be cast in molten metal, into a metal plate that could finally be used to print a page of newsprint on rolls of paper with ink.

The pressmen who worked there had mangled fingers and ink stains like semipermanent tattoos on their arms. They'd be smoking cigarettes from the moment they came to work until they left at 4:00 a.m., working late every day to print the news. In the lunchroom, they'd munch on hot dogs and donuts, the smell of sauerkraut as thick as the ink stains on the walls.

I imagine Gutenberg's workshop to be somewhat similar, with ink-stained and metal-scarred men working in dark rooms, sharing their lunches at a dark table in the back, drinking beer together. And maybe there's a dog or two in the corner, nuzzling at some grunt sleeping off the beer he had for lunch. The workshop would have been smoky from lampblack, boiling linseed oil, and cauldrons of molten lead.

You would hear the sounds of the press as it was squeezed, like a grape press to make wine; you'd hear the groans of grunts as they turned the screws of the printing presses, the creaking of wet wood against metal.

Scraps of books and Bibles would litter the floor, along with pages from calendars showing the best time to do bloodlettings or letters from the Pope printed to rally support against the Turks. There would be splurts of hardened metal on the floor and rows of metal slugs. With a word as soft and slippery as "slug," it's hard to imagine them being made out of metal, but a *slug* is a line of type made from copper, all the letters neatly arranged and ready to be inked for printing.

In my mind, the workshop would be divided into sections for casting molten metal, for pressing it into paper with ink, and for the racks of moveable type that could be maneuvered into place to set each line by laborious line for whatever was being printed at the time. In Gutenberg's day, it was too expensive to print an entire page of a book from one copper plate, which is how they did it at the newspaper when I was a child.

Copper was only affordable enough for Gutenberg to do one line of type at a time. He could set a line and then disassemble the letters and words once that line was printed. If he needed to print more of the same Bibles or books, they would have to be hand-set from scratch, page by page, all over again. But it was the most he could afford.

It was a dark and secretive workshop, just as secretive as any tech company today, for fear of outsiders stealing this brilliant idea. Even in the 1450s, secrecy was paramount. It was an age before patent law, and innovators had no other way to protect themselves besides confidentiality. There was talk at the time that the English and Dutch were developing their own printing presses, and Gutenberg had to be careful. So all the Germans in his workshop huddled together and kept their knowledge and books secret within the fortress of its walls.

Amazon, Apple, and Google are a bit like medieval fortresses in their own ways. They're secretive like China or Japan before they were opened up to Westerners, or like Tibet or Mecca, closed to foreigners, with rare exceptions like Sir Richard Burton, an explorer who dressed like a local and sneaked inside with his binoculars and surveying rods. In a way, it's appropriate to speak of Amazon and other ebook companies in a medieval sense, because although ebooks seem so advanced, we're really just emerging from the Dark Ages of reading today.

Gutenberg was just as obsessive as Steve Jobs of Apple or Jeff Bezos of

Amazon. He is known to have spent months worrying about how many lines of text should be printed on one page and varying the number to find the optimal balance between cost and aesthetics. By increasing the number of lines, he could reduce the number of pages that needed to be printed, but this made the book more difficult to read.

Interestingly, I've seen the same situation play out in Amazon's conference rooms. I've been in meetings with Jeff and his vice presidents where he obsessed about the number of lines that would appear on the Kindle screen. I've seen his 3:00 a.m. emails after those meetings. I've seen his mind wriggle and squirm just as obsessively as Gutenberg's, and about the same feature. It's as if we had to reinvent printing during the ebook revolution, and men like Jeff Bezos, Steve Jobs, and Eric Schmidt were Gutenberg's ghosts, reincarnated hundreds of years later.

It's still amazing to me that a billionaire like Jeff—with an enormous business empire and a company that makes rocketships, no less—would take hours out of his life to obsess about line spacing, of all things! But attention to detail matters. Revolutionary innovations and products live or die by such obsession—and I believe Jeff wanted Kindle to be his legacy to history. He wanted it to succeed.

Printing as we know it eventually emerged from the Dark Ages. The actual quantity of books produced during the early years was relatively small—but with every decade afterward, books became cheaper as new technologies were introduced. Mezzotint. Offset printing. Lithographs. Electric typesetting. The mass-market paperback. And as literacy around the world increased decade by decade, more and more books came to be printed because there was an audience for them.

With the advent of internet technologies like eBay and Amazon, it was possible by the late 1990s to find and acquire almost any book ever written (for a price, anyway). By the turn of the twenty-first century, we were practically flooded with printed books. Once the bulwarks of wealth and prestige, once gilded and bound in fine leather and placed behind glass cabinets in showy libraries and drawing rooms, books were now cheap commodities. You could go to any used bookstore on a weekend and see racks and racks of books sitting forlornly under an awning—a sad sight for a book lover.

As a culture, we're still very bookish, very literate, even though books are no longer the entertainment medium of choice they once were. And while it's easier to collapse at the end of a hard day on your soft couch in front of your TV or laptop to watch your favorite show, there's still a place for books in our lives, because they are the rawest and truest form for telling stories and collecting, analyzing, and communicating information and ideas. The beauty of books is that you approach them at your own pace. Not only can you read at your own speed, but you also can skip from section to section, nonlinearly.

Of course, books have their limitations. They're heavy. They're hard to lug around on a vacation or pack in boxes whenever you move into a new home. Books are cumbersome, and it's hard to find what you're looking for in them. They can get out of date quickly. They age and mildew, rot and crumble.

Those of a future generation will one day look back on printed books with the same benign and befuddled expressions that we use when we look at floppy disks or those colossal IBM mainframes with spinning reels of tape that you see in the background of the villain's lair in James Bond movies. Books are bulky, and an individual book doesn't hold much data compared to what an e-reader can hold.

Please don't misunderstand me. I'm a book lover: some of my best friends are books. But I see the limitations of books, and I see ebooks as their natural continuation. And yes, this means that one of the challenges we're going to have as ebook readers is to accept that reading is a technology-based experience. That means the culture of reading will evolve and change like all technologies do. This might seem troubling to some, but remember that print technology has also evolved over the centuries. It simply had a 500-year head start, and there aren't many evolutions left for it.

By the time you and I started reading as kids, print was basically done evolving. But we're still on the rapid exponential rise of technology's evolution for ebooks. Also, as an insider in the publishing and retail worlds, I can tell you that you're going to start seeing far more ebooks and fewer print books. Readers are migrating to digital, and ebooks are a more attractive financial proposition for publishers; the economics are simply better.

Print books will, of course, still be published, but primarily for blockbusters, the kinds of books that will get lots of press, lots of advertising. Of course, there will also still be an attractive market for print books as collectibles, whether they're antiquarian books or special-edition commemorative hardcovers. But ebooks will rule the day, and when people a few years from now talk about "books," what they'll really be referring to are ebooks, not print books. Eventually the "e" will be dropped, and books will be assumed to be digital, just as most music is now digital; after all, we don't refer to music as e-music.

The future of books is fraught with possibilities and dangers. With ebooks, we're no longer reading on paper but on eInk or LCD screens, and although each type of screen has its own technology, behind every e-reader is a hard drive of some sort, something that stores the books you read.

Hard drives are the new clay tablets for books. The reason we love them so much is that they're so cheap to manufacture, these thin wafers of silicon and circuitry that are often made without any moving parts. Hard drives are ridiculously convenient, and our civilization rests on them; the web itself is supported by air-conditioned data centers all around the globe, vast buildings where hives of hard drives hum away.

Convenient, yes, but also prone to failure. The average hard drive has a 25 percent chance of dying after three years, so there are employees at Google's and Amazon's data centers who do nothing all day long but trundle down corridors with carts of replacement drives. Hard drives are convenient as long as you have anything that resembles a computer or cell phone or e-reader, but with the consequence that any content on them is likely to disappear fast as once-hot electrons grow cool, as magnetic fields flicker and fade.

At least clay tablets were given the dignity of turning into dust, but when ebooks die, it's without a sigh. We can still archaeologically excavate former libraries and palaces in the Middle East and turn up tablets and parchment, but no one will be able to take a shovel to Apple's data center in North Carolina thousands of years in the future and unearth all of these servers, power them up again, and recover all the ebooks they once held.

Indeed, book technology has reversed itself. Clay tablets, once durable but inconvenient, have been replaced by hard drives, which are highly convenient but very fragile. And though our words are now more widespread than ever, they barely have the lifespan of a hamster or a gerbil. They're short-lived unless they're constantly restored and backed up to new hard drives, to new computers. Ours is a culture that dances on the edge of ephemerality.

If our servers slept for too long or if we left our iPads unplugged for too long, we'd wake up like Rip Van Winkle to find all of our book culture erased. And in the ultimate progression, if you look at this curve of decaying durability and increasing convenience over time, the inevitable trajectory is for our words to last as long as the twenty-four-hour life of the mayfly, to be as ephemeral as a June bug's jitter. Someday, our ephemeral but instant thoughts themselves will be beamed in a quickenth of a second from brain to brain in some ultimate evolution of print technology. But will they last beyond that fraction of an instant?

I think we've made a proverbial pact with the devil in digitizing our words. And digitization raises questions: since we've traded durability for culture, what happens if there are massive failures in our culture's data centers? What happens if ebooks are one day wiped out? Viruses can now target nuclear power plants, so is it not conceivable that viruses could be developed to destroy ebooks?

Bookmark: Reading in Bed

All of our brains are alike in some fundamental ways. We can all get into a book, whether it's in print or digital format. We can all zone out and ignore the fact that the book is on an eInk screen or is bound into pages with glue that smells like fish vomit. We can ignore these things if it's a really entrancing book.

One of the times I enjoy reading the most is at night. I love reading in that golden hour before bed, where you're able to put away your computer, your cell phone, anything distracting that keeps your mind unduly alert. And instead, you allow yourself to relax into a damn fine book. It doesn't need to be a printed book; an ebook works just as well. Because as far as the brain's concerned, the experience of reading an ebook can be the same as reading in print, in that you're no longer aware of the medium itself—most of the time, anyway.

At night, though, when you're trying to sleep, using LCD screens or backlit e-readers can actually zap the production of melatonin in your body, keeping you awake longer and degrading the quality of the sleep you do eventually get. Thus, I rarely read a tablet before bed, preferring an eInk reader instead.

If you're using a glare-free e-reader that doesn't zap you with light, reading in bed is even more of a pleasure than before. E-readers are lighter than most books, and you can annotate them without having to fumble in the dark for a pen. You're absorbed into the book, and you can fall asleep with the ebook in your hands.

In fact, this is my test of how well a new e-reader is built. If I can fall asleep with it while I'm reading at night—feeling my eyelids growing heavy and closing with that gluey cotton-candy sensation of sleep overcoming me, feeling the e-reader slip out of my hands and onto my bed—then it worked. The ebook was able to transport me into that stuporous yet sublimely sensible state of pre-sleep, the state in which I have my best ideas, when I'm able to let go of the rational and uninhibit the intuitive. I don't care whether you're reading in bed or on an airplane or on a train; you can still fall asleep with a really good digital book, if you allow yourself the leisure to do so.

But that's just me; I like reading in bed. Some people love to read with cigarettes and endless cups of coffee in a Manhattan diner. Some people love to read on their computers, endlessly following one link to the next, one website to another. Some people like to read at work between bites of their sandwiches or salads, their office doors closed during their lunch breaks, indulging in some downtime, some time for their mind, time for a book. Some people don't like to read, but there's no room for them in this book.

Perhaps my favorite kind of reading happens at palmy beach resorts. There's always a stack of used paperbacks at a beach resort. With dog-eared and salt-encrusted pages brittle from exposure to the sun and scores of readers, these books are left out at night in a makeshift library, home perhaps to hermit crabs and wasps. But as e-reader costs come down, I actually expect that we'll start seeing a bunch of left-behind e-readers instead of paperbacks at beach resorts.

Five or so years from now, you'll be using someone's discarded Nook or Kindle to read in a hammock in the jasmine-scented wind, swinging in late-afternoon sunlight to a gentle breeze with a gentle book. Their eInk screens will be sun stained, and the saltwater won't do them any good. The power chargers will go missing, and the wires will get frazzled and frayed. But on the plus side, you'll be able to surf the web—albeit more slowly than you'd like on your laptop—while listening to tiki music, your stomach rumbling in anticipation of dinner and a frozen margarita.

How about you? Do you have a favorite sunny nook or diner where you prefer to read? Do you prefer reading in train stations, on airplanes, or sitting cross-legged in the aisles of an expansive, book-filled library?

http://jasonmerkoski.com/eb/2.html

The Origin of Ebooks

The ebook revolution started in 2003 when the vice president of research for Sony spent a week trapped in an ice cave after a skiing accident. He had a broken leg and was forced to drink his own urine until he was rescued. As he stared at his cell phone, hoping for a signal, he thought, "If only I had a book about wilderness survival on a cell phone or some other device that I could pull up right away." And that's when the idea for Sony's first e-reader hit him…

That's not what happened, of course.

The origin of ebooks has more to do with technological inevitability. It's like the story of printing itself—Johannes Gutenberg was in the right place at the right time and knew enough about minting and metallurgy to make it happen. The founders of the ebook revolution were also each in the right place at the right time.

And because the creation of ebooks involved technology, there's no one origin story, no pat answer for how they came to be. Like all stories, that of the ebook revolution has as many twisted roots and beginnings as there are people who were part of the story.

You could, for example, trace the roots to the invention of electrophoretic ink. Xerox discovered eInk in the late 1970s but then mothballed their invention when they couldn't figure out how to sell it. In the late 1990s, pioneers in Cambridge, Massachusetts, rediscovered eInk and improved it, and it's partly due to their efforts that we have ebooks now.

Even before the dot-com boom got underway, there were predictions that eInk would be commercially ready as early as 2003, and anyone could have been the first to launch an e-reader. But Sony was in the right place at the right time.

Like most companies that sell consumer electronics, Sony has to invent a major new product every year to stay competitive, and eInk technology had matured to such a point by 2003 that Sony decided to use it to manufacture an e-reader. Insiders at Sony have told me that the ebook team's budget was tiny compared to the budget for Sony's TV division. In fact, the ebook team had to salvage parts from the Sony Walkman to make the first Sony e-reader. So even if the eInk technology was mature, the Sony e-reader was expensive and only available to the top echelon of the reading population in Japan, where it was first released.

Of course, if an e-reader were to launch anywhere first, it would have to be Japan! The Japanese are among the most techno-literate of all cultures. I've traveled to Japan a lot, and I always smile when I encounter new tech marvels there, feeling rather like a bearded German in a fur cap from Gutenberg's time transplanted to modern-day Tokyo.

I've gone to the Sony showroom to see the display of next year's toys and gadgets, the talking dogs and robot servants, things that you won't see on shelves in the United States for months or maybe years. I remember once standing outside a Tokyo convenience store for five minutes, trying to figure out the way inside, before noticing a subtle hand-plate recessed into the wall to touch and open the door. Even Japanese toilet technology is extreme—the Japanese are truly techno-obsessed.

But they weren't obsessed enough for ebooks to take off when the Sony e-reader launched in 2004. The Japanese language is a challenging one, and the Sony device didn't do a great job of rendering content in that language. Plus, there simply weren't many ebooks to buy. So Sony took the product to the United States in 2006 and launched a revamped e-reader here.

Now, Amazon had time to watch Sony and learn from its mistakes. Amazon followed what its competitor did best by using eInk displays and basic metaphors like bookmarking and page turns. But Amazon had also accumulated ten years of book knowledge and was sitting on millions of books in its fulfillment centers.

Almost half of all the books bought in North America are sold through Amazon.com. It represents the single biggest chunk of the bookselling pie. Unlike Sony, Amazon had customer brand loyalty for its books, because it had started by selling books online and had worked hard to build its brand. Any early adopter of Amazon.com in the 1990s was given all sorts of freebies with every order, like T-shirts and coffee mugs.

Amazon succeeded at ebooks in part because Kindle was a new business line and capitalized on the company's success in the bookselling market. Kindle also didn't have to worry about turning an immediate profit, unlike the Sony eReader. The Kindle organization was in some ways a startup within Amazon and benefited from Jeff Bezos's venture capital infusions, long-range vision, and full support. Amazon was in digital media for the long haul. Since nearly all of its sales still came from physical media when the Kindle project started, Amazon clearly knew it would have to take a long-term view of digital media, like ebooks, to keep growing.

The other reason Amazon succeeded was because it focused on creating an easy, seamless customer experience. Consider trying to use Sony's e-reader when it was first launched in the United States. To read a book, you had to:

1. Download an application to your PC.
2. Find a book you wanted.
3. Log in.
4. Purchase the book.
5. Authorize your computer to use the Sony device.
6. Download more software from Adobe.
7. Authorize yourself with Adobe.
8. Go back to your library and try to download the book.
9. Synchronize the book with the device (assuming you have the right cable to plug into your computer).
10. Wait a few minutes for it to (hopefully) finish copying.
11. Disconnect the device from your PC. Now you could read the book.

Whew! In comparison, the Kindle is simple: you go to the online store; you find a book you want; you buy it with one click; and then it downloads immediately and you start reading. No hassle; nothing to it. Delivering content to Kindles is that easy because each device has a built-in cell phone that is always on and connected to the national network.

There have been two great inventions so far in the twenty-first century. The iPhone is one of them. And even if I didn't work for Amazon, I'd say that the Kindle is the second. Ebooks as we know them finally took off because of the Kindle's embedded cell phone and free data plan. Without a connection to the cloud, I don't think ebooks would have become mainstream.

A network connection goes beyond making it easier to get content onto an e-reader. It also makes it possible for you to instantly read an ebook loaned to you by a friend or to freely sample the first chapter of millions of ebooks. In addition, the network lets you easily redownload books you've previously purchased. You can even accidentally break an e-reader and redownload all your old books onto a new e-reader with no hassle or technical wizardry. The network acts like a safety net for all of a Kindle's content.

Technically speaking, we could have had ebooks as early as the 1970s. That's when people started digitizing the first ebooks. In fact, I can imagine librarians in their bell-bottom jeans and with their "Whip Inflation Now" pins archiving books onto microfiche. I can imagine a digital revolution for ebooks starting back then. As I mentioned earlier, scientists at Xerox discovered eInk in the 1970s. They could have developed e-reader hardware using electro-phosphorescent displays. Instead of Amazon digitizing the world's content, the Library of Congress could have been doing it. They could have started digitizing their holdings in the late 1970s, Xerox could have made the device, and the teletype network could have been used to distribute content. Though the process would be considered slow by today's standards, an average ebook could be transmitted over teletype in about ninety minutes.

But that's a future that never was.

Sony started the ebook revolution. But if Kindle was to do for reading what the iPod did for music and what TiVo did for digital television, Amazon needed to make a device that not only used the cell phone network, but also took advantage of new game-changing technologies such as eInk, which is touchy and temperamental.

I'm not going to try to fully explain how eInk works, with its vocabulary of ghosting and quantum mechanical waveforms. You see, eInk is actually based on quantum mechanics. I studied quantum mechanics at MIT, and I still don't fully understand eInk!

Perhaps the most appropriate metaphor for something like eInk, which is at once scientific and magical, is that of the Magic 8 Ball. You shake it up and ask a question, and a ghostly white answer mysteriously floats to the surface. That's similar to how eInk works. A bright-white particle, usually made from titanium rust, is electrically charged and floats in black ink. But instead of shaking the ball to get the white to rise to the surface, you apply an electrical charge.

If you do this enough times, with hundreds of thousands of tiny bits of titanium rust, you basically get the modern eInk screen. The ink is black, and the charged particles are white, producing the two simplest colors on an eInk screen. To get shades of gray, you apply a quick pulse of electricity, just enough to attract some of the particles, but not all of them.

Arthur C. Clarke could have been describing eInk when he said that "any sufficiently advanced technology is indistinguishable from magic." True, eInk is complex, but it requires very little power. In fact, an eInk-based device can function on as little as one charge a month. This was the game-changing technological leap that allowed the ebook revolution to start.

Bookmark: Annotations

I don't annotate my books. Personally, I think that defiles the printed page. But I know that some people see annotations as a cherished way of life, a way of reconnecting with themselves as they were across the span of years. These people can look at their books and see what they highlighted years earlier with their pencils or fluorescent markers.

All e-readers let you annotate to your heart's content. You can underline whatever you want, and your annotations and highlights will, of course, follow you from device to device. That is, assuming you buy devices from the same manufacturer.

I think Amazon will support its own ecosystem for handling annotations, as will Sony. But there's no interoperability yet for annotations among different devices, and there may never be. For the next ten years, your annotations will probably be tied to your choice of ebook retailer. Once you choose a retailer, you're going to be more likely to stick with it, because you're going to want your annotations, highlights, and all the books that you already purchased to follow you around as part of your ongoing library.

But what happens decades from now if people want to see what you wrote in your books—perhaps because of scholarly, archival, or genealogical interests? If you're not around anymore, or your account with Amazon or Apple is closed, your annotations will be gone.

That's sad, because annotations add lasting value in helping to understand a person's path through life. One of my favorite books is a very dense volume called *The Road to Xanadu*, which was written in 1927 about Samuel Taylor Coleridge's mental life. Its author, John Livingston Lowes, analyzed all the books that were in Coleridge's library and books he borrowed from friends, as well as annotations he made in those books and in his journals, and pieced together how Coleridge came up with every word in every line in just one of his poems. The 600-page book attempts to explain exactly how his imagination worked for that one particular poem. That kind of literary detective work simply wouldn't be possible without annotations left behind by the original author.

No one I know is planning an archive service for annotations. It's a potential startup opportunity, although a very niche one. Perhaps such a startup will preserve all our ephemeral electronic annotations for posterity. While the current crop of e-readers offers the ability to add annotations, those notes are often a lot more free-form and messier than text entries on a printed page.

For instance, my mom's cookbook is stained a hundred hues of saffron and turmeric. It's speckled with tomato paste from numerous attempts to make pasta sauce and splattered with bits of molten butter from exploding Yorkshire puddings. Every page in the cookbook is a food-encrusted testament to meals we once had.

No ebook can capture the history of so many Thanksgivings and Sunday brunches like my mom's cookbook can. It's like a combination of a scratch-and-sniff book and a time machine. The food stains themselves are palpable annotations of former meals, and I've got to tell you, I still use a print cookbook for my own cooking. It's better to have cookie batter on your cookbook than on your iPad.

Other annotations are more wordlike, but they capture you as you once were. On my writing desk I have a Wolf Scout book I used as a kid, from the time I was eight years old. It has a list of activities inside it, such as "List the ways you can save water" or "Name four kinds of books that interest you." Free-form fields follow each activity, filled with the answers in my own handwriting. Below the form, you can see my mother's signature and the date. So not only do I know to the day when I first learned to tie an overhand knot or put a Band-Aid on my finger, or learned to use a pair of pliers or notify the police of subversive Communist activity in the neighborhood (I grew up at the end of the Cold War after all), but I also have annotations of most of these events in my own handwriting.

My handwriting in the book is labored, cursive, and bold; graphologists could look at my annotations and perhaps learn something about me. But they'd never learn anything beyond the factual from a sterile ebook annotation. There are paint splatters in this old Scout book, mud smudges, and decals from a Pinewood Derby racer that I built with my dad. How could they possibly fit inside an ebook, unless future e-readers allow you to insert photos inside them? (Let it be known that I never did get my merit badge in spotting Communists.)

Is there a viable future for annotations? Perhaps. I see a glimpse of it in a recently launched web service called ReadSocial. This web-based system lets readers not only annotate a given ebook, but also comment on one another's annotations. Best of all, it works for a variety of different ebook formats, and it's as easy to use as logging on to Twitter or Facebook.

By working across multiple ebook vendors and being brand neutral, ReadSocial (or one of its competitors) has the potential to become the de facto annotation engine for ebooks. Such a service may not preserve decals from my Pinewood Derby race car or smells from my mom's cookbook or, for that matter, annotations from any print book, but it may pave the way toward creating compelling conversations in the margins of ebooks.

And after all, isn't that what we're looking for? To find a kindred spirit in the pages of a book—the voice of the author or perhaps another reader—to carry on a conversation with? In this spirit, why not connect with others right now? Click on this link to meet a kindred book lover through the conversation about this chapter online.

http://jasonmerkoski.com/eb/3.html

Launching the Kindle

Working at Amazon was like taking a step back in time to Seattle's pioneer roots, back when Seattle was the gateway to the Yukon gold rush. Working on Kindle was like living in the Wild West.

For projects that broke new ground, like Kindle, there didn't seem to be any law, any sheriff, or any real consequences for making wrong decisions, because nobody knew the right ones. People seemed to wear their six-shooters out in the open, taking potshots at one another while hiding behind Donkey Kong machines. When vice presidents argued in the hallways, trigger fingers twitching, I could almost imagine a tumbleweed blowing between them.

It was also impossible to tell reality from fiction. No outsiders had seen the Kindle because it was created in a perfect vacuum from the very beginning. Everyone was trying to do the right thing, and no ideas were off the table. Nothing was too strange to consider. People who thought fast often got their way and ruled the day. It was an early Wild West of ideas and innovation. It was crazy and anarchic, and I liked it.

Download a copy of *The Diamond Age* by Neal Stephenson. It's the book that all of Kindle's hardware code names came from. The book is

about a character named Fiona and her "illustrated primer," a machine designed to look like a book but with links to all libraries, all TV shows, and all human knowledge. (Jeff originally wanted the Kindle code names to come from *Star Trek*, since he's such a Trekkie, but more literate minds prevailed.) The book is a treasure trove of other code names for Kindle hardware: Nell, Miranda, and Turing.

So the first time I got a Kindle, it wasn't called a Kindle but a "Fiona."

Though primitive by today's standards, my original Kindle—one of the first Fionas made for select Amazon employees—still works like a charm. True, my Fiona is turning the yellow-gray color of smokers' teeth, the same way that once white yesteryear computers start to turn an upsetting beige. But it still works, even though it's been manhandled and chucked many times into my backpack, tossed into many suitcases for trans-Atlantic flights, and left on my truck's dashboard in the sun for months. And once while walking through Cupertino, California—a city where everyone drives—I got hit by a car while crossing the street, because nobody expects pedestrians in the heart of Silicon Valley. I fell and sprained my arm. But even though my Fiona clattered to the street and got run over by one of the car's wheels, it still works as great as always.

Needless to say, I love my Kindle.

My original Kindle job had me creating and managing the ebook conversion process—the messy method by which print books are turned into digital ones.

When thinking about how ebooks are created, it's best to envision a sausage factory. Meat comes in one end, machinery packages it, and sausage comes out the other end. At the ebook factory, you start in the front with books from publishers. They're chopped up, reassembled and packaged, and finally made available for sale in digital form.

Most ebooks are created using a digital copy of the physical book, usually in PDF format. PDF files have a fixed layout, which means they're formatted in the way they're supposed to appear on a printed page. However, ebooks need to be reflowable, which means that if you change the font size on the ebook, the words and sentences and paragraphs should be reformatted so that the words wrap around properly in the paragraph. You can't do this well with PDFs.

To make a PDF into a reflowable ebook, publishers usually use a conversion house. Such companies, in turn, use a combination of software and workers overseas. Many of the conversion houses use people in India or China, or sometimes more exotic places like Sierra Leone or Madagascar or the Philippines. They usually work in a large warehouse or an old factory, with cubicles running from one end of the factory to the other on multiple floors.

Elbow to elbow, the workers stare at words on the screen all day, reading ebooks. They remove page numbers, reformat the ebooks to make them reflowable, and skim through them afterward to make sure no paragraphs or illustrations from the originals were lost during the process.

But not all books are in PDF format; some only exist in print. More brutal methods are often needed to digitize such books. As part of my job, I got to watch as workers destroyed print books to turn them into ebooks. Pages had to be removed from books so they could be scanned and digitized. As a book lover, I was horrified. To remove the pages of the book, workers would hack the spines off with knives like they were whacking their way through the jungle with machetes. Once their content was scanned, those pages would be tossed into a Dumpster at the end of every shift.

It was destructive, and the books could never be recovered afterward. The ebook revolution was bloodless, in the sense that there were no human casualties. But if books could bleed, you'd find their graveyards overseas. You'd find burial pits, unmarked graves, and hundreds of thousands of casualties.

But all this was needed to launch the Kindle; we couldn't just launch a hardware product without any ebooks to read. Without ebooks, the Fiona device would have been just an expensive paperweight.

You see, we needed both the ebooks and the hardware for the Kindle flywheel.

Many people in dot-com and tech companies think in terms of "flywheels," but most nontechnical people don't know what that means. It probably sounds like lots of flies strung up to a mill wheel, slowly turning it to crush wheat into flour.

In tech terms, a *flywheel* is something that builds up energy as it spins. The goal is to get it spinning faster and faster, however you can.

The faster it spins, the more energy you have (or in business terms, the more money you have). The Kindle flywheel, for example, might start with launching an e-reader into the marketplace with a small number of ebooks. People buy the device, and then they use it to buy ebooks. The profit from both can be used to build an improved e-reader, which can be sold more cheaply, which then means more people will buy it and consequently buy more ebooks, the profits of which can then go back into building even better, even cheaper Kindles. With every push the flywheel gets, the faster it spins and the more powerful it becomes.

The Kindle flywheel started spinning fast as the Kindle business grew. And in true Amazon tradition, the business was run with metrics, with meetings called "deep dives" where the team would dive into spreadsheets. Amazon is a highly numerate culture. The numerically literate seemed to do well there, because they could mentally pivot rows and columns of spreadsheets and crunch numbers on the fly.

During a deep dive, you let go of preconceived notions and think logically. You look at data—instead of doing a technical hand-wave, you speak to the specifics. In Amazon's deep-dive culture, facts are preferred to opinions. Deep dives are like science experiments, and you approach them with a hypothesis you want to prove. If your hypothesis is disproven, then you come up with a new hypothesis, run tests to gather data, and analyze data to prove or disprove the new hypothesis.

Most of the engineers at Amazon dreaded these deep dives because they had to put on something formal, like a button-up shirt and a pair of jeans with a belt. Amazon isn't a formal place: a J. Crew shirt and Dockers are as formal as it gets. But still, for engineers, even wearing these is an affront against nature, a blasphemous abomination out of a Dungeons & Dragons game or an accursed H. P. Lovecraft story.

In one of my first meetings with Jeff Bezos, we were doing a deep dive on ebook content and what it looked like on the Kindle. We sat and used our Kindles as customers might. In some ways this was like the first digital book club; we were mostly silent, just reading on our Kindles. Sometimes we would annotate content or buy a new book—anything to test all the features.

At one point, Jeff's Kindle must have crashed, because it became unresponsive. The room had been silent for a while because we were all absorbed in our books. Then out of nowhere, Jeff exclaimed: "I'm hung! I'm hung!" I looked up with a surprised grin on my face, but Jeff was unaware of his double entendre.

All the others in the room were actively trying to stifle their laughs. There was a little bit of hero worship at Amazon. Now, I admire anyone who runs a bookstore, so I can't help but admire Jeff Bezos. Not only does he run the world's biggest bookstore, but heck, he has his own rocketship company too. But some of my colleagues took admiration to a whole new level.

I don't think anyone at Amazon deliberately shaved their heads bald to look like him, but people would be in a Jeff meeting and come out afterward and rave about Jeff's stories, how he laughed, or a savage insight he had. People would find out about the books he was reading and read them too. (During the Kindle years, *The Black Swan: The Impact of the Highly Improbable* was popular among the Jeffnosanti, although a book he read on the history of tungsten was slightly less popular.) People routinely lionized Jeff for how much money he had and his high IQ. So they certainly did not want to look like they were laughing at him or criticizing his ideas.

Let's face it: we all contributed to Kindle, but Jeff was the visionary, and digital books will be his legacy. True, there were other digital book pioneers. Heck, I was one of them. I made the first modern ebook in 1999, and I invented my fair share of Kindle features. And I wasn't alone; we all invented Kindle in our own ways. None of us who toiled in the Kindle workshops were flunkies. We were all colorful characters, innovators, and pioneers.

But only Jeff had the vision and the millions of dollars in seed capital to start Kindle. And trust me, it took a lot of capital, considering the salaries and stock grants for the employees the first few years, as well as all the R&D and acquisitions and startups he had to fund. Jeff not only saw the dream; he also made sure the dream happened, at great financial risk.

So as difficult as our challenges could be, life at Amazon felt like we were creating something revolutionary, and we had the financial means to do it. We were like techie versions of the early workers who toiled in Gutenberg's workshop.

Life in the Kindle offices in those early days was like working in an alternate, over-caffeinated, sugar-high universe. And I loved it. The offices were loud, with the sounds of BlackBerries and pagers going off. The building shook every ten minutes as a streetcar rumbled past, and the hum of a microwave melting someone's leftover Indian dinner filled the air at lunchtime. Inevitably the cries of an engineer shouting at the top of his lungs would emerge from a conference room, along with the pounding of his fists against the whiteboard walls.

In the kitchen you would find occasional stacks of Top Pot donuts, local Seattle fritters that tasted like they'd been deep-fried in nothing but pure sugar, cocaine, and aspirin. You'd also find the remains of catered breakfasts or lunches that senior management would put out in a kitchen for anyone else to have when they were done eating, like lords of the manor throwing their serfs an occasional bone to nibble on.

Like most technology companies, Kindle had lots of beer, usually on Friday afternoons. People would often bring in six-packs, open them near someone's desk, and stand around and talk at the end of a long day. Some crazy conversation would emerge, full of crazy, hypothetical what-ifs like: "If you could suspend a killer whale from a rope and suspend a tiger from a rope and let them attack each other, who would win?"

"Decoration" was hardly the word for what you'd see in the warren of cubicles we inhabited. If you walked among them, you would find Amazon-issued Magic 8 Balls, humming computers, Kindles connected to power cables, teetering printouts of architecture diagrams or spreadsheets, posters from *Battlestar Galactica* on how to spot Cylons, tipsy engineers still arguing about whether the killer whale would get the tiger, discarded Kindle boxes being used to prop up Foosball tables, and an arcade-style, fully functioning Donkey Kong game I could never beat.

Amazon was, in short, a bit of a sloppy Seattle dot-com—but one with billion-dollar revenues and razor-thin profit margins. Those thin margins meant that we had to stay focused on launching Kindle, without distraction.

Secrecy was important in the early days of Kindle. We weren't allowed to take our Kindles home or show them to our families or get caught using them in public, out of fear that someone outside of Amazon would see the Kindle and leak information to blogs or newspapers.

But with this secrecy came a great feeling of pride and privilege. I felt like one of the first people to use an iPod, years before anyone else even knew it existed. The Kindle was a secret I couldn't share with anyone, not even my family!

Until the Kindle launched, the only other place on the planet that knew about it was Lab126 in Cupertino, California, where the Kindle hardware was designed.

In the very early days of Kindle, when its eInk screen was just a gleam in Jeff Bezos's eye, Amazon was smart enough to realize it had never done manufacturing before. It was great at website sales, but it had no expertise in making hardware. Jeff decided it would be best to spin up a new organization solely responsible for this.

The name Lab126 came from a technical kind of pun. Amazon already had its "A to Z" development center in Palo Alto, which developed technologies like the A9 search engine that Amazon uses. Jeff wanted Lab126 to be a research facility—hence the "Lab" part of its name.

As for the "126" part, well, you have to realize that there was never a Lab125 or a Lab124, just like there was only ever a Preparation H, never a G or an F. The "126" part stems from the fact that "A" is the first letter of the alphabet and "Z" is the 26th, a techno-geeky homage to the "A to Z" development center. Jeff liked his geeky in-jokes—you could have heard his laugh a mile away when they came up with that name.

To attract and retain the best hardware engineers, Lab126 would be located not in Seattle but in Silicon Valley. The Lab126 offices were originally in a mini-mall across the street from a music studio and a slightly sleazy jewelry store. But due to the number of new hires, the lab quickly grew out of its old space and moved to Cupertino, right in the heart of the Valley. Moving to Cupertino put them in the big leagues—they were now in the same city as Apple.

After a year running Kindle's ebook software team, I was asked to take the lead in launching Kindle as its program manager. That meant I had to know everything about the Kindle hardware, so I started flying down to Lab126 on a regular basis. I went to bridge the gaping cultural chasm between Amazon and Lab126. Everyone in Cupertino understood hardware, and everyone in Seattle understood the web, but neither understood the other. Amazon understood web services; Cupertino understood consumer electronics.

And the combination of the two: ebooks? It was new territory for everyone. Almost no one in the company had exactly the right set of qualifications to help Lab126 and Amazon speak to and understand each other, with one exception: me. I used to work at Motorola making cell phones and internet routers, so I could speak the language of hardware people, but I had also built websites for companies like Home Depot and Walmart, so I could speak the language used at Amazon.

On visiting Lab126 for the first time, what you'd notice would be the stark contrast in floor layout between Amazon's offices and Lab126's. Amazon has a messy organic layout. All the floors are open, with people at desks sitting side by side in a vast room without walls, like a Southeast Asian call center or some fly-by-night dot-com's tech support division. The Lab126 offices resemble a printed circuit board, in keeping perhaps with the mentality of a hardware engineering company. All the cubicles and hallways are aligned at right angles, with efficient pathways in between. Being at Lab126 was like being on the circuit board inside a Kindle.

Much as I wanted to read on my Kindle while flying on a plane every week between Lab126 and Amazon, I couldn't because Kindle was still such a secret. I couldn't even bring the Kindle through airport security, in case the security personnel needed to examine it. Plus, I feared that journalists or competitors might see my Kindle in the few seconds that it would be in plain view.

I spent two years traveling back and forth between Lab126 and Amazon. When you're a kid, the years have a way of passing quickly.

All you can seem to remember when you look back are summer nights, fireflies, and snowball fights. The same was true of me with Kindle. When I look back at the years leading up to the Kindle launch, it's like I was a kid, moving happily from one day to the next, one challenge to the next.

One of the challenges required Jeff's personal attention and had to do with the Kindle ebook format. Nobody else on the Kindle team believed it was important enough to merit his attention, but I did, so I set up a meeting with Jeff to discuss it. (I suspect that the days when someone can set up a meeting willy-nilly with Jeff are over, now that Kindle has grown so large.)

Now, just because you had set up a meeting with Jeff didn't mean it would actually take place. To get to Jeff's office, you had to get past his executive assistants. They have offices of their own, and in a Kafkaesque way, you'd have to talk your way past the first executive assistant to see the next one and then talk your way past her to make your way to the third assistant, and so on. Eventually you got to Jeff's office, where you'd probably find him gone and realize that they'd neglected to say he was out for the day.

The day of the meeting, I made it to Jeff's office a little early, before he had arrived from another meeting elsewhere. I looked through his windows and tried to understand the way he saw things. He had a telescope in his office and pictures of his kids on the wall. It was a small office, actually, dominated by a giant work desk with tidy stacks of papers.

I imagined him looking out through his telescope at his far-flung workers, spread out as they were through Seattle in different office buildings, and I imagined him perhaps aiming his telescope at his fulfillment centers in Kentucky or Nevada, yearning to see the incessant shipments of everything from books to Beanie Babies, DVDs to diapers.

Protected by his executive assistants and sequestered in a tower in Amazon's headquarters, Jeff's office was a little like a walled garden. It's an appropriate metaphor, because what Jeff and I discussed that day, and on days and weeks to follow, had to do with Kindle's own walled garden.

When you're reading about companies like Amazon and Apple, you often come across the "walled garden" metaphor. I want to explain it to you with a visual metaphor, because I'm a visual guy.

Imagine the wall of a medieval fortress. There might even be a moat around it. It's a tall wall, made of stone—a wall to keep the enemy out. There's one way into and out of the fortress, and that's over a drawbridge that comes clanking down to let you across the moat, through a hole in the wall, and into the city inside. You can think of the city as being everything good that the wall is supposed to be protecting—all the people and gardens inside. This wall protects you from the dragons outside, from the Vandals and Huns and would-be conquerors.

In tech terms, the *walled garden* is the arrangement of software and hardware and file format that makes it almost impossible to get to what's inside unless you go over the drawbridge, the officially sanctioned way in.

Look at the iPod. It relies on a proprietary format, a proprietary way of getting content into and out of the device. And yet it's successful because the walled garden is tended so carefully.

Amazon has a similar walled garden for the Kindle. The only way you can buy a book and read it on the Kindle, according to Amazon's walled garden approach, is to buy the book from the Kindle store. Are there other ways of reading a book on a Kindle? Yes, but they're equivalent to the Vandals and Huns laying siege to the city by running ladders up its ramparts and then climbing those ladders with axes and grappling irons. In modern tech terms, this kind of attack is piracy. Or if not outright piracy, it's that gray area related to digital rights management (called DRM)—the restrictions used to keep people from copying or sharing ebooks for free.

DRM works against most would-be pirates because it's often too difficult and exasperating to break. Difficult, but never impossible. It's a game of cat and mouse, and there's always a genius who outsmarts the current DRM that's out on the market. And then the software people at Amazon and Apple and elsewhere respond with patches and updates to make their walls more secure. Apple, for example, releases about ten updates a year to its iTunes software, and most of them include anti-piracy measures.

As ethical readers, you and I don't need to worry much about what DRM means, and it's not likely to affect us. But because occasional readers do try to pirate ebooks, we're all penalized by the increased cost of ebooks and the inconvenience in copying them to other devices. That process should be easy, but often it's painstaking. I think everyone agrees that it's sad that we have to live in a world with DRM, but it's a consequence of the technical nature of ebooks.

Likewise, there's another technicality with ebooks called "file format" that we don't have to worry about with printed books. There's only one format for a printed book, and that's paper. You can pick up any printed book and read it, as long as you know the language. The format of the book is no barrier to reading.

But imagine having to wear special glasses to read books by different publishers. Imagine you needed one pair of glasses to read Random House books and another to read Simon & Schuster books. Each pair of glasses would be sensitive to the invisible inks each publisher used. Well, that's what it's like with ebooks now.

Amazon has its own ebook format, and Adobe makes another ebook format called *ePub*. There are many formats on the market. If you live in Japan and want to read ebooks, for example, you have two incompatible ebook formats to choose from.

Formats make things difficult. There's no way I can take a book I bought for my Kindle and copy it onto a Sony device—not unless I use some technical wizardry, some illegal tools that can be downloaded from the shady side of the internet. And most consumers aren't going to learn how to use these obscure wizard's tools. Like you, they'll be confronted with a choice of e-readers, which locks you into the format of the books you're able to read. And once you're locked into the format, you're locked into what kinds of books you can read. You may find that a book you want to read is only available for Kindle, but if you have another device, then you can't purchase the book until it eventually becomes available on that device.

The Kindle has its own proprietary format. And it's an old format, one that dates back to the 1990s and applications written for PDAs. Now, I worked at Amazon, and I know the Kindle format inside and out. I couldn't have told Jeff this at the time—but as much as I hate to say this, I believe the Kindle file format was limited and made for poorer-quality ebooks.

Here's how to think about ebook file formats: think about their fidelity to printed books. When we speak of music, we often speak in terms of lo-fi or hi-fi, that is, low or high fidelity. The same terms are appropriate for ebooks. Think of a print book as the gold standard for quality. A book in the Kindle format would be able to reproduce most of the text (though not all the accent marks and sometimes obscure symbols) and is often able to reproduce the margins or the page breaks in the print book.

I estimate the Kindle format to have achieved something like 50 percent fidelity compared to print. It's on the lo-fi side. But I think formats that launched in the years after Kindle, such as the one being used in both the Nook and iPad, are more hi-fi, because they allow designers to do typographically compelling high-design flourishes, as well as embed fonts and complex equations into the ebook. These formats approach 90 percent of print fidelity.

I'm a book lover, and I cared a lot about improving the file format when I was on the Kindle team. But to Jeff and others, file format was just one of many issues that needed to be taken into consideration in launching Kindle. And besides, in the early days, most people on the Kindle team didn't worry about these lo-fi and hi-fi details, because Kindle was targeted at readers buying genre fiction like romance books and sci-fi and bestsellers. Even in print, these kinds of books aren't stylistically nuanced.

I dealt with crises of all stripes and sizes as we prepared to launch Kindle, but all the issues eventually got solved, one by one.

Before I knew it, there was just one day left before Kindle launched.

We don't know what it was like for Gutenberg in the hours before he unveiled his Bible and the secrecy was finally lifted. Until then, was he furtive, fearful that any secret would be stolen and copied? We don't know how he or his workers felt. Sure, we know that pies were introduced in the 1450s, and we can imagine Gutenberg going outside with his workers that day and serving them celebratory slices of quail pie and glasses of plum gin or something special from his larder.

Some of his workers were no doubt hungover from the night before, drunk in a corner and being licked by the dogs after celebrating their victory, but maybe others could see how important the printed book would be. Because truly, Gutenberg had launched something at once commonplace and innovative—a humble Bible, but one set in beautiful, printed type. He unwittingly launched the Protestant Reformation, as well as a shift in reading so profound that we're feeling aftershocks of the original tremor even now, centuries later.

Five hundred years later, on the eve of the ebook revolution, I settled in for sleep the night before we launched the Kindle. But sleep was impossible; there was the nagging worry that I had surely forgotten someone or something important. I kept getting out of bed to check my email. I finally managed to get an hour's rest before being awakened by a team in India looking for help with some last-minute problems.

After helping them, I stayed awake in bed with prelaunch insomnia, looking out through the window and thinking. Tomorrow, once Kindle was launched, things would never be the same again for anyone. Amazon had a lot of power, and ebooks would surely capture people's imaginations.

I stayed awake through the early morning hours of November 19, 2007, wondering about the Kindle. What would ebooks mean for literacy, for reading, for the book itself? Would the Kindle hasten the decline of the book—a decline that had started with radio and movies and had accelerated with TV and video games and the internet—or would it instead revitalize books and breathe fresh life into them?

Such questions still keep me up at night. I have answers for some of my old questions, but now I struggle with new ones. On the morning of the Kindle launch, I looked through the bedroom window until I had to get dressed and go to work at 4:00 a.m. There was a rare break in the gloom-clouds over Seattle, and I could see a few stars, bright enough to be planets or maybe omens.

The next few hours saw me running the show in Seattle, while Jeff Bezos was on stage in New York announcing the Kindle. The launch was timed to the minute; I had a clipboard and a stopwatch. I was like the mission-controller in the movie version of *Apollo 13*, the one with the sweater around his shoulders who made sure each team was "go" for launch.

We didn't want anyone saying, "Houston, we've had a problem," which is why the launch was scripted and tested in advance. The script was flawless. It was a dream launch. We got the store and services running at almost the exact moment when Jeff said, "Introducing Amazon Kindle," in front of thousands of reporters and bloggers.

And then, Kindle was live.

Everyone in the Amazon offices in Seattle, sugar-addled since 4:00 a.m., started cheering.

We have the founders of Napster to thank for the widespread adoption of digital music, and we have Netflix to thank for the adoption of digital video. But the future owes digital books to Jeff Bezos.

Jeff is a simple man. His front teeth are a bit chipped from when he grinds them together, and as the years passed, he seemed to grow thinner, his snazzy blue suit slowly engulfing him. What hair Jeff had when I first met him gradually disappeared entirely. He has a great laugh, an infectious laugh. It makes you smile, as all great laughs do. As Jeff stood in New York about to announce the Kindle to the world, I could only imagine what he must have been feeling.

This was the moment Jeff had been waiting for since 2004. As he said in his press event that day, "We did a number of things that make the experience of discovering new reading material, getting that material into your hands, and reading seem like magic."

And he was right: it really was like magic. As magical as books themselves.

The sorcerers behind the magic of products like the Kindle are the product managers. If you're lucky as a product manager, you'll have the time to dream up new ideas, but if not, you'll be handed ideas from other executives and told to figure out how to make them happen. Some product managers are more expert than others, more visionary. At Amazon, for example, the CEO was the ultimate product manager.

And although such product managers are possessed of genius, there are two other secrets to their success. For one, they're poised like spiders in the centers of their webs of information, and they feed on this network of information. They know more than anyone else in their web and can use this information to further their own projects. Secondly, they have the enlightened autonomy to pursue their goals—something that can't be done in politics or academia. These are enlightened capitalists for whom even their boards of directors and shareholders will often look the other way, trusting in their long-range plans and their long-range genius.

Three years earlier, Jeff had embarked on the tough challenge of inventing a new kind of book, a new kind of reading experience. But now, as we launched our first product, not only could we all finally read in public with our Kindles, since it was no longer a secret, but we also could introduce others to the joys of ebooks. We could change the lives of our customers by making reading more immediate and more featureful. We could continue innovating, using the original Kindle as a launch platform. We could continue adding improvements to a fundamental human experience, one that hadn't changed in more than five hundred years. We were giving customers something they never asked for and delighting them with something at once strange, magical, and uplifting.

As for me, I could finally call my family and tell them what I was working on. For the last few years, I couldn't say because Kindle was confidential, so my parents thought I was working for the FBI! I was excited and humbled. I rode the bus home and proudly read my Kindle and showed it off to everyone—although I was so exhausted that I don't think I was able to read more than a page. I was temporarily relieved, but I knew that there'd be even harder work in the months and years ahead—not just for me or for Amazon, but for the billion-dollar book industry.

Bookmark: Knapsacks, Book Bags, and Baggage

Our Stone Age ancestors developed an innovation that I doubt few of us today could replicate, alone in the wilderness: the simple pot.

Whether it held water, seeds, or honey, I think the pot was the single greatest invention of the Stone Age. Before its invention, people most likely had to live closer to rivers or try to carry water with their hands, a futile task. Containers like the humble pot allowed people to spread geographically, to move and transfer goods and objects easily, and to improve the quality of their lives in a game-changing way. I think the ability to conceptualize and enclose volume in a man-made artifact is one of the keys to civilization.

The high-tech equivalent of the humble pot is the information cloud.

We don't know where the cloud is taking us as a society. It's something like a magic carpet, and we're aloft on it, flying above everything, uncertain of our destination. The cloud is in essence a container for digital goods, and it's already revolutionized the way we store those goods. It's a clever way of enclosing yet more content in a much smaller area. The cloud is a giant pot with near-infinite volume and near-zero size. I'll expand on this subject in the chapter "Our Books Are Moving to the Cloud," but for now, I'll note that because of the cloud, we no longer have to haul ebooks or information with us as we travel.

That makes satchels, book bags, and hand baggage increasingly useless as we adopt ebooks.

As a kid, I would manhandle an enormous book bag in school every day. I never had time to run back to my locker and replace books between classes, so I carried my full day's allotment of books with me to all the classes I attended. After four years of this in junior high and another four years in high school, my shoulders were unusually well developed for a skinny, nerdy guy. But it was frustrating, tiresome work. I needed to buy a new book bag every few months. And every year, we would be inspected for scoliosis in gym class, no doubt partly because of all the books we had to haul, crushing our spines into sad, deformed springs.

Luckily for kids and their back doctors, this is no longer necessary.

And on adopting digital books, you no longer need to haul boxes of books with you every time you move to a different home. Gone are the days of duct-taping shoddy cardboard boxes from U-Haul or liquor stores and still watching your books explode onto the sidewalk when movers accidentally drop the over-heavy boxes. As the heir to the Stone Age pot, the cloud makes moving easier for those of us with large holdings of books.

A digital book weighs less than the whisker of a fly. So there's no strain with the digital. You don't have to haul digital books in cardboard boxes or book bags, so digital books are easy on the shoulders, and on the eye. But clearly, I'm a believer in the digital. Are there drawbacks to ebooks, in this sense? Absolutely. The sheer massiveness and weight of books adds a kind of gravitas to a home. Books in a home say that someone literate lives there, someone with specific sensibilities and tastes. A home with fully digitized music and ebooks and other media seems barren to me, like a minimalist Bauhaus detention cell, someplace unfit for friends and family. But that's me. What do you think of books as decorations or as hefty physical objects to be lugged about?

http://jasonmerkoski.com/eb/4.html

Improving Perfection: Launching the Kindle2

Improving the Kindle meant more than making better hardware, although I didn't realize that immediately.

As a program manager, I got to fly into any building, any country, and do whatever it took to get my product shipped. A part of the job was making sure that people were on schedule, but another part was more punitive, requiring me to check out their dirty laundry. I had to be the eyes and ears of the Kindle executive team. And to do this, I had to know more about the Kindle than almost anyone except Jeff Bezos.

Being Kindle's program manager let me see how decisions were made all across the Kindle organization. I participated in meetings with teams all over the globe, as well as with the vice presidents and Jeff in Seattle. I had an opportunity to see and influence what was happening with Kindle hardware and ebooks in this position, and by being with Kindle leaders, I learned a lot about Kindle and the Amazon business. I could see the personalities that shaped Kindle.

For a year and a half, I found myself flying to Silicon Valley every week, because Lab126 was where Kindle2 was being built.

The Kindle2 was an improvement in design compared to the original. It was lighter, and the eInk was crisper, with more shades of gray and more nuance. The device fit better into your hand while reading, and it had some cool features, like being able to read books out loud to you. It was also much cheaper, even though it had more features.

With Kindle2, almost everything was reinvented from scratch. Even things as seemingly insignificant as the box it shipped in.

The original Kindle package was a very maximalist presentation. It was designed to look like a hefty white book. You opened the book and found the Kindle inside, as well as its leatherette holder and a special sleeve for the power supply, all neatly arranged. On the outside of the package, and imprinted in rubber on the underside of the Kindle, you'd see a wonderful explosion of symbols, like someone had thrown a hand grenade into a type foundry.

But for the second Kindle, the package got reduced to a simple cardboard box with no markings at all on the outside, nothing to indicate there was a Kindle inside. And yet when you opened it, you'd find a beautiful Kindle sitting on a plastic tray, like a pearl in an oyster on the half shell. The packaging was simple and functional. In fact, with its nested layers of plastic, culminating in a strange dishlike tray, the Kindle2 packaging had all the aesthetic charm of a TV dinner.

Amazon moved from an ornate package design to a simple cardboard box that could be sent by UPS or FedEx and left on your porch without anyone knowing what was inside it, the same kind of box that could be stocked on the shelves at Best Buy or Target. It was practical, but soulless.

Although this packaging was more cost-effective, there was no artistry to it. I'm a big believer that industrial design is a sign of the times, and I'm not alone in this. Andy Warhol would look at department stores like they were museums. I love looking back at 1920s typewriter tins and 1930s talcum powder cans, industrial designs from eras when they still showed zeppelins and aeroplanes flying overhead as signs of their times.

If someone looks back a hundred years from now at our current industrial designs, they'll perhaps see our culture as being obsessed with digging through layers of plastic and cardboard to get at the pricey prize inside. They'll perhaps misjudge us and accuse us of not having any artistic inclinations. But they shouldn't be too harsh on us just because the CEOs of our largest tech companies were frugal. Because inside these boxes were some of the most incredible devices in history.

Almost everything improved with the next-generation Kindle. By the time we were finished, the Kindle2 was truly an incredible device, with features we were sure would amaze the next generation of ebook readers. But the way there was paved with endless reinventions and trials that left all of us sleepless and stressed. As the head of it all, as we moved ever closer to launch, I started to sense myself being pulled closer each day to a breaking point I had never felt before.

The day we finally launched Kindle2 was almost a sleepwalking dream for me. I remember Seattle being shut down by a snowstorm that day, and I remember how buses careened into one another. One slid off a bridge and into Puget Sound. Cars can't drive up the steep Seattle hills in snow, so many were simply abandoned until the snow melted.

It was February 2009, a rough time to launch. I came in at 4:00 a.m. again and saw starlight again through holes in the clouds. After the launch, I was numb to news about the number of Kindles we sold. Twenty hours later, I climbed back into bed and slept for a week.

In fits of wakefulness, I thought about how Kindle lacked nuance, style, fonts, and things like multimedia. How great it would be if you could have a book about the history of music with actual musical excerpts! These seemed like great ideas to me, but I wondered if they were a bit too ambitious for Kindle. Because by now, Kindle's success made new ideas paradoxically difficult, as if everyone was walking around on stiletto heels on a glass floor, careful not to run, not wanting to take the wrong risks.

I also realized that there was no outreach to the outside world—to publishers especially. I thought Kindle should have evangelists, like Guy Kawasaki once was for Apple, out there in the magazines and on the trade show floors talking about Kindle products. Not just as a paid shill, but as someone who used the products and believed in them with a fervor that approached religious fundamentalism. And that's when the energy started to come back to me.

I realized that it was one thing to improve the Kindle as a device, but another thing entirely to improve the content. Over the last year and a half of effort, nothing had been done to differentiate the ebooks themselves. They were still the same as before. No worse, but no better.

The only category of books that I think the second-generation

Kindle improved on was pornography, of all things. This is because the number of shades of gray on the Kindle2 doubled. Porn sells well in any format, whether magazine or book, but it sells especially well in ebook form. Amazon prefers not to sell pornography, but that doesn't stop many users from buying it elsewhere and loading it onto their Kindles. With the Kindle, you could download pornography to your device and read it anywhere, even on a subway, without anyone guessing that you were not reading the latest bestseller. Digital books excel at protecting a reader's privacy while he or she reads. And in this same sense of protecting privacy, digital books are the best thing that ever happened to pornography, with the possible exception of the brown paper bag.

The drawback with pornography is that images can look awful on eInk, no matter how much you dither with them. The original Kindle's 2-bit eInk screens, for example, only had four shades of gray, and of these, one was white and one was black, which didn't leave much for nuance. Whether you're trying to render a picture of the sky or of a woman's thigh, it's hard to get pornography to look good with only four colors. Depending on your stance on pornography, that's either something wrong with eInk or something you're glad e-readers don't do well. Even with the Kindle2's sixteen colors, digital pornography still sucked, although it did improve somewhat.

Clearly, there was potential for improvements in content beyond pornography. There was a whole universe of books to adapt to e-reading—including atlases, dictionaries, comics, travel guides, and textbooks!

Shortly after the Kindle2 launch, I talked to Kindle's senior management and then took on the role of Kindle's technology evangelist. I would be half evangelist and half product manager and focus on ebooks alone. A product manager is something like a practical futurist, someone who can think nine months into the future and see a product through from inception to launch. I would be able to dream big and make long-range improvements.

I was refreshed and revitalized, ready for a new chapter of my life at Amazon. I was on a plane every week as Amazon's first technology evangelist. I would meet with publishers in their midtown Manhattan offices to explore new ebook ideas together. Then I'd be off to India or the Philippines to see how conversion houses were making ebooks and to tell them some of what I'd learned during my time at Amazon, feeding them bits of information that would make them work better and faster and cheaper for us. I was doing my small part with each player in this ebook ecosystem to move it forward and to find ways that publishers could spend less and convert more, so readers could have more ebooks to enjoy.

I saw colossal, warehouse-sized machines that stripped books of their spines in seconds, like wood chippers for books, but that were as precise as a doctor's scalpel. At a technology park in India, I also saw an experimental array of quarter-million-dollar machines that were like animatronic spiders. They were used for nondestructive scanning, the high-end way to digitize content—unlike the cheaper method of hacking pages with machetes. The machine lifted a book and carefully turned its pages one by one so they could be photographed and digitized. Those animatronic spiders were so delicate that I would have trusted them to hold a baby and change its diaper.

Being an evangelist gave me a chance to engage with publishers worldwide, and I got to see the scramble firsthand as publishers adjusted to digital books. Some publishers reacted better than others; some, in fact, were downright revolutionary.

Ultimately, I think everyone who worked in those early years of ebooks was changed by the experience. We weren't working just for paychecks. We were learning and growing. We changed from one month to the next, sort of like taking a paintbrush and a bucket of water and drawing your self-portrait on a hot sidewalk. You'd maybe be able to sketch half of your face before the water you'd already painted would start to fade and evaporate, so you'd never quite be finished.

We're all sidewalk portraits painted with water on a hot summer's afternoon. And there's a holy fervor and zeal than you can see in the eyes of the ebook revolutionaries who are working as insiders, whether they work at the publishers or the retailers or as independent software vendors and sideline pundits. It would be one thing perhaps if we were merely part of the MP3 or digital video revolution or part of TV in its early test-pattern era. But (and you know this already) there's something sacred about books. They're humanity's lifeblood, these inky words and smudges that make their way into our minds. Ultimately, books are a small but essential part of the human condition. They are tapestries of birdsong, magic, and intrigue, in equal parts.

As an evangelist, I was interacting with publishers and ebook revolutionaries outside of Amazon. I was moving the ebook revolution forward by improving ebook content. I was venturing beyond Amazon's walled garden to plant seeds like a Johnny Appleseed for ebooks, never quite certain how these seeds would grow but certain they needed to be planted. And then those seeds would grow and bear fruit that would find its way back to Amazon and to the Kindles and ebooks I loved. This new role was a first for me and for Amazon, with its highly secretive culture. For Amazon, this move was revolutionary.

Bookmark: Burning Books

On July 12, 1562, Diego de Landa, the Bishop of Yucatan, started a horrific bonfire. Hundreds of Mayan scrolls were tossed into the fire, as well as thousands of sacred images. Diego de Landa believed himself to be in the moral right, having found what he called "superstition and lies of the devil" in the books. He had gained the trust of the Mayans, gained access to their sacred books, but then with the might of the Spanish conquistadors behind him, he burned them all. Only three full scrolls of the formerly vast Mayan empire remain now, plus charred portions of a fourth.

The Nazis too are known to have burned books. Jewish and "degenerate" books—including volumes by Albert Einstein and Ernest Hemingway—were raided from libraries by Nazi youths and consigned to flames. At least 18,000 distinct titles were identified as officially objectionable, and untold hundreds of thousands of copies were burned in well-attended public events.

Book burning has historically been a tool used by tyrants in authority to penalize or marginalize detractors. Do you think America was more enlightened? Not really. Even though we value free speech in America, we have at times taken a tyrannical approach. During the McCarthy era of the early 1950s, it was decided that "material by any controversial persons, Communists, fellow travelers, etc." would be removed from libraries and burned. In fact, this was enacted by presidential decree.

It's harder to burn ebooks.

Burning an e-reader will cause you to choke from the fumes, so don't do it. And while digital book burning won't happen, a more subtle version might arise. The handful of retailers who control the distribution of digital books could choose not to make one or more books available for any number of reasons.

Consider the time, shortly before the iPad launched in 2010, when Amazon decided to yank the "buy buttons" from books and ebooks published by Macmillan, one of the top U.S. publishers, to protest new pricing terms Macmillan wanted. Amazon removed tens of thousands of books in all.

It was one of the brazen moves Amazon sometimes makes. Pulling the "buy button" from items in the store means that it's not possible to click on any button to actually order a given book shown on the Amazon web page. You can see the book—it's tantalizingly close—but you can't buy it. As long as the buy buttons are gone, orders can't be placed.

It's a money-losing proposition for Amazon and any business partner it decides to yank the buy buttons from, but contractually, it's something Amazon is allowed to do. But why would the online retailer want to do that? It's like Amazon is shooting itself in the foot. Perhaps Amazon had previously shot itself in the foot so many times that it thought it had bulletproof shoes. Or enough scar tissue not to mind.

Yanking the buy button is a punitive gesture that Amazon has been known to pull with publishers, like a tyrannical Byzantine emperor who holds ultimate sway over his court. It's a powerful threat in business negotiations. But Macmillan wasn't a mere vassal to some king's court. That publisher is an empire unto itself in the publishing world. The move to yank books backfired when publishers became enraged and retaliated as a unified group. Ultimately, Amazon needed the books and the support of publishers and its customers, so the company backed down.

Some choices are tough, but leaders are judged by the decisions they make when given tough choices. I believe the Amazon leaders made a mistake. An ethical retailer has a social contract to uphold with its consumers. It's not appropriate for a retailer to yank or censor content based only on its internal machinations, its policies for better profit margins.

Thankfully, I believe this example shows the power of public outrage to enact change. It's possible to shame a corporation that has done something wrong or, at the very least, to make a company aware that it should have been more careful about its actions. The same public outrage was hurled at Apple when it released a "Baby Shaker" app that rewarded users who could shake a virtual baby to death. Developed by another company, this iPhone app is grisly and should never have passed Apple's QA standards. Mercifully, public outcry caused it to be pulled from the app store in less than a day.

No company has perfect QA policies or editorial standards for what content to shelve in its store. Companies need to listen to consumers, read what people post on product reviews, and monitor the blogosphere. But reciprocally, companies need to have strong enough standards in place to avoid smear campaigns and acts of undeserved bullying. Knowing when to remove or reinstate content requires an ethical balance and strong sensibilities.

It's a tough editorial choice: though a given book may be objectionable, where do we draw the line when it comes to free speech? And more importantly, who is drawing the line? What moral or literary sensibilities do the executives of Amazon have? What about the retailers at Barnes & Noble or Google or Apple? You have to ask yourself whether you trust these men (because they are mostly men—and mostly white men, at that). Do you trust them to make decisions for you on what books you're permitted to buy?

http://jasonmerkoski.com/eb/5.html

The First Competitors

You can create an innovative breakthrough, but you can't own it forever. Eventually, competitors come out of the woodwork, challenging you with similar or sometimes advanced versions of the very innovation you crafted. For Amazon, the first major competition came from an old rival, a company that Amazon was used to competing with in books. But it was a company that, back then, would have seemed most unlikely to make a tech-product marvel. Yet in November 2009, that's exactly what it did.

Los Angeles is all sunshine and short sleeves. It's still got a 1960s design, like it was influenced by *The Jetsons*, but with more palm trees and fewer spaceships. It's got atomic roadside diners and terrific tangled, spangled freeway sprawl. It's got the best mom-and-pop taco shacks in the most unlikely of places, like wedged between Laundromats and exotic pet stores in strip malls.

I'm at one of those strip malls on a long layover from a flight, visiting a Barnes & Noble store. I've been sitting here for a few hours watching people. I've been watching the kiosk where a saleslady named Bettina is showing off Barnes & Noble's new Nook e-reader. A few people come every now and then to look at these Nooks. More often than not, people come up to ask her where the bathroom is or what time the store closes, like she works the information booth. The Nooks aren't exactly selling like hotcakes.

I go over to her and show an interest in the Nook. To torment her a little, I keep calling it a Kindle. "What can these Kindles do?" I ask. She laughs, explains, and walks me through a demo. I tell her that it would be nice if some sort of sticker on books inside the store would show if they're available on the Nook. Something on the print book's front cover, right there on the retail shelves. It's the sort of thing that Barnes & Noble can do but Amazon can't because it doesn't have a physical presence.

After a few minutes, some sweatered, grandmotherly looking men come over to look at the Nook, and so does a woman with so many facial piercings that she'd probably set off a metal detector. I slowly drift away.

I love real-world retail. It connects you right to your customers, without the web as a nameless, anonymizing barrier. Bookselling as we know it emerged at the end of the Roman Republic, around 50 BC, at a time before publishers existed. Retailers would contract directly with scribes, copyists, and authors. Then they'd create lists of books for sale and post them for customers to see outside their stores on Rome's winding side streets. Bookselling as we know it grew more complex with the proliferation of separate roles for authors and publishers and retailers, as well as the advent of copyright and of securing rights for publication, and the explosion of mail-order and online commerce.

Though I've worked in online retail for two decades, I still never get tired of looking at bookstores. I'm a bookstore tourist whose first vacation priority on arriving in a new city is to check out the local independent bookstores. And I have a special place in my heart for Barnes & Noble, the biggest of the retail bookstores.

They're sharp on the ebooks side, as well. Out of all the retailers who sell dedicated e-readers, they're the most innovative. They were the first to release new book-reading features and to innovate on the hardware side. They were the first to have touch-sensitive eInk screens. They innovated digital book lending for swapping books between friends. Heck, if you're in one of their stores, you can read any Nook book for free for an hour or so. They totally get the social experience of books in the way that it crosses over from the real world to the digital.

They can innovate so fast because they're not burdened with their own R&D group. Instead, they use a company called Inventec, a sort of hired gun in the world of R&D. It's a kind of Lab126 that hires itself out to the highest bidder. By outsourcing the nuts and bolts of their product development, Barnes & Noble can focus on innovation.

Their Nooks are downright futuristic too. When I first got my own Nook, I was just as perplexed as everyone because it had a big eInk screen for reading and a thin color screen at the bottom for navigation. The day I opened my Nook for the first time, I was sprawled out on my living room floor like a child opening a birthday present. (Okay, a birthday present I had bought for myself.) The Nook's dual screen is clever and innovative, even if it is neurocognitively jarring. (When you get confused by all the screens you have to navigate, that can take you out of the reading experience.)

One of the reasons that Barnes & Noble makes such innovative devices is because they don't have to worry about building their own operating system, unlike Apple and Amazon. Those two companies are slowed down by the boatloads of engineers who tweak and tune and build an operating system from scratch. Barnes & Noble simply uses Google's free Android operating system, which lets the retailer put its engineers on other projects to make e-reading even better.

Barnes & Noble is innovative with the software as well as hardware. For example, the Nook was the first e-reader with a game platform. So you have to give credit to every engineer and director at Barnes & Noble for what they've done.

Even more so, what they did with interactive children's ebooks threw the publishing industry for a loop. For the first time, you could actually play with an ebook. You could touch an elephant to hear it bellow, or you could become a character in the book. And it's not hard to extrapolate from interactive children's books to interactive books for adults or readers of any age.

If you ask me, though, in spite of such interactivity, ebooks aren't ready yet for children. I think a children's book should be sacred and sensual, an inviting canvas for the imagination that can be colored in with crayons. For children, words are already puzzles. They're strange glyphs that children need to decode as they yearn with outstretched fingers for fluency in their language and to grow up into readers. Games can be distractions from that process.

Most publishers agree, and I think they're right to move slowly on children's ebooks, because being a digital native may have long-term consequences related to learning how to read. We're in danger of rushing a whole generation of children into something unplanned and unexpected.

And while I like the occasional TV show, I still look back at my childhood with some resentment because the television was often my babysitter. I was raised by Buck Rogers and Oscar the Grouch and geriatric game-show hosts like Bob Barker. And I can still quote the price of Cocoa Puffs from the 1980s, thanks to *The Price Is Right*. Digital books, like television and other media, are best meant for those Pandoras who've already opened their boxes and know what demons to expect inside.

That said, I applaud the Nook team for inventing interactive ebooks. It was a bold, innovative move. And one that Apple and then Amazon were soon to copy. Likewise, when Nook introduced ebook lending, the other retailers were swift to add that feature.

Ebook innovation is a game of cat and mouse. Unfortunately, one of the drawbacks of this game is that it becomes all-consuming—and innovation becomes harder to do when you're trying to keep up with competition. When Apple launched a tablet, Amazon had to follow suit, even though it undoubtedly had other features on the drawing board, innovations that wouldn't be launched until at least one other retailer had launched them.

I think some competition is healthy, because it forces an evolutionary Darwinism of features: if a feature is successful, it will be copied. But untested features languish in unread business requirement documents, and resources that would have gone into building those features get redirected into keeping up with the Joneses.

Amazon is winning the ebook revolution, but it may lose the war. Competitors like Barnes & Noble and Apple have successfully blurred the lines and proven that they can provide a great media experience, so Amazon's brand matters less in the eyes of readers now. Any tactical advantage Amazon has is primarily related to its deep ties with publishers, ties that are much deeper than those of other retailers, except maybe Barnes & Noble.

The revolution started with one clunky, four-hundred-dollar device with four shades of gray that could only hold a hundred books, but the war is about all media now, about the convergence of books and audio and video. The war is on as different retailers compete for your attention. Books were once hugely popular, but they have been relegated to a small slice of the media pie. And though book media is still a billion-dollar industry, it's becoming outranked by TV and movies and audio and video games in per capita media consumption.

A 2010 Nielsen survey of American households showed that books account for only 3 percent of an average family's monthly discretionary spending, while music accounts for 5 percent, video games 9 percent, and videos a whopping 29 percent. There's no room for niche players to succeed at just selling books, which is why the digital retailers are getting into the game with all kinds of media. And now that ebook content is being sold at commodity prices, the true differentiator will ultimately be in the reading experience itself.

The winner of this war won't be decided by generals with scale models of battleships and airplanes and tanks on a simulated table. No, it will be decided by designers, by user-interface artists, by people who connect to the humanistic spirit that flourished in the Renaissance as print books gained in popularity. The Renaissance saw the rise of readable fonts, innovations in binding and page layout, and the placement of illustrations. And typographers always experiment, whether with the more lavish encrustations of the Art Nouveau period or the German grid style that emerged in modern times.

In the end, design matters.

Spend a weekend in Los Angeles, and then spend a weekend in Seattle, and ask yourself which city you'd rather live in. Seattle started out as a logging town and as a gateway to further riches in the Yukon.

Its roots are founded on the exploitation of resources, as if there's an infinite supply of trees to chop or gold to mine. Historically, Seattle is a city that has drawn hard-core, hard-boiled businesspeople. That's why you see the likes of Microsoft and Amazon and Boeing around Seattle. Frankly, it wasn't the most auspicious place to start a venture that would revolutionize books.

Places like New York and Los Angeles are still rooted in the arts. New York has theater and publishing and advertising, and L.A. was founded on Hollywood, the movies. You don't get that artistic sensibility in Seattle, and you can tell by looking at the current Kindle and all the knockoffs that copy its design. Bullish as I am about ebooks, something is missing, and this flaw is perpetuated by the fact that all the e-readers are made in Silicon Valley. Apple, Amazon, and Barnes & Noble all have designers in Silicon Valley because that's where the technical talent is. But what you don't get with this technical talent is an artistic, book-oriented design.

As consumers and readers, we're not dummies. We don't want an impoverished reading experience. We don't want a cracked plastic case and a blurry screen—which, sadly, is what many e-readers offered, especially in the boom years between 2007 and 2012 when everyone seemed to be trying to sell a budget e-reader. For good or for bad, we define ourselves in many ways by the gadgets we use and the clothes we wear. We don't want to surround ourselves with cheap products. Nobody really aspires to that. We also don't want to pay for a diamond-encrusted e-reader. We don't need bling; we just need to feel like the design speaks to us.

That's the genius of print book covers. There's a reason why print book covers evolved to a highly specialized, soulful art form. They add very little to the cost of a book, and yet they make reading a vibrant, colorful experience. When you think back to a book you've read, you'll often remember the cover before you remember any words or ideas. As designers re-embrace the original strengths of print books, I think we'll see more book-oriented themes in future e-readers.

Eventually, the line between print and digital will blur and finally vanish. Ebooks borrow from print books now, in terms of their design metaphors. They copy bookmarks and annotations, as well as the concepts of turning the page and of page numbers themselves—even though page numbers don't even make sense in an ebook.

What's a page number? What's a page, if you can dynamically change the font size or the font? What's a page, if you have a game embedded in the book and the game spans many levels? These design metaphors are yesteryear bolt-ons from physical to digital. But there's an opportunity to reinvent the digital reading experience while keeping the best parts of print.

Companies with more humanistic sensibilities than Amazon will win the e-reader war by making the experience more human, more engaging. Children's ebooks should be playful and adult ebooks thoughtful, soulful, or entertaining. Companies should create opportunities for interesting, unexpected experiences to happen. Perhaps digital insects scuttle across the page if you've had the book open for too long without turning the page. Perhaps in a thriller, as you read the ebook, you're startled by the unexpected sound of a gunshot when you turn the page to a crucial passage. Though this can't easily be done in hardware, you can create an engaging experience in software and make it soulful instead of awful.

Let's face it: there's still something emotionally bereft about a Nook or a Kindle. Perhaps over time the industrial design will become more human, more like the "illustrated primer" described in Neal Stephenson's *The Diamond Age*. Or like the book Penny used in the *Inspector Gadget* TV series, a digital book with actual pages that could be turned. Better design will be part of the rebirth of reading. But to get there, we must be as ready to innovate in design and soul as we are in technology and cost.

The company that does this best is Apple. They blew away everyone's preconceptions about e-readers when they launched the iPad.

The iPad story is a great one, but I wasn't a part of it. I didn't have to put in the grueling hours or countless meetings on product development. It's one of those stories that we get to read and enjoy. One of those stories where you say that the author did an amazing job. And Steve Jobs did.

Apple understands a lot, including great product design. The iPad is a multifunction device, unlike other e-readers that are dedicated to reading. Dedicated e-readers are as sharp as steak knives in doing what they're supposed to do, which is let you read books. The iPad is more like a Swiss Army knife—it can cut the steak and uncork a wine bottle, and there's even a toothpick to use when you're done eating! It's got it all.

Sure, it's got its flaws. For example, there's the headache the iPad gives you when you try to read in direct sunlight, since it doesn't have a nonreflective eInk screen. But overall, Apple did an amazing job in creating a product that actually feels like a book. The iPad has the same heft as a book. It's got the same-sized screen as the average printed book, and it's as responsive as a book when you turn the page. You don't have to wait half a second like you do with eInk readers. It's truly, as Apple is fond of saying, a magical device. A device consumers love.

That should be no surprise, because as one of America's favorite companies, Apple has some of the most famous lovemarks of all.

A "lovemark" is the concept that a brand isn't enough. That brands are dead and products are commodities, so to make a product succeed, it needs the love that comes with fads and the respect that comes with established brands. If you've got this love and respect, you've got a lovemark—something that combines intimacy and mystery and sensuality.

A great example of a lovemark is the Swiss Army knife. Every time you open it, you find a new tool or screwdriver or spoon or toothpick or who knows what inside, prompting surprise and delight and gratification. It's gratifying in the same way the Kindle was when it first came out. With a Kindle, you could download a book in less than sixty seconds. (It still stuns me today that you can do that.) Even the name "Kindle" connotes something mysterious. What's more intimate than the experience of curling up with a book to read? Amazon got a lot right with the Kindle that made it into a lovemark.

And what's even more amazing is that Amazon created a lovemark by focusing on the product, not the ad campaign.

The first round of Kindle commercials emphasized the lovemark, imaginatively making books come alive, in a way that was at once amateurish and approachable. But later commercials have seen a return to Amazon's retail roots. In one, a man and a woman are sitting by a pool in Las Vegas. He's reading a print book and she's reading a digital book on her Kindle, and she cheerfully explains what the Kindle does and how it's cheaper than a pair of designer sunglasses, pulling no punches when it comes to commoditizing the Kindle. There's no lovemarking here.

Such commercials treat Kindles as mere commodities with price points that serve utilitarian needs. Admittedly, it's the kind of commercial you expect from a wholesale retailer of goods, but it's not the kind of commercial that speaks to the soulful, mysterious aspect of books themselves. I suspect this disparity turns some would-be consumers away from Kindle.

Barnes & Noble's commercials for the Nook, on the other hand, are intimate and sensuous. One follows a beautiful little girl through childhood and adolescence and adulthood, and you get into her mind as she reads. The commercial stays true to the way that reading simultaneously transports you to new places and comforts you where you are now. At the same time, the commercial factually communicates that the Nook is an e-reader with a wireless connection and all sorts of content available—something the original Kindle commercials never quite conveyed.

The Nook is an under-appreciated genius of a lovemark. The team at Barnes & Noble got a lot right with the Nook, and from a lovemark perspective, I think they created a more intimate product than any other dedicated e-reader. The rubber back behind the Nook is soft and pliable—not hard metal like the later Kindles—making it sensual and intimate. Barnes & Noble also recreated the engraved faces of famous authors from their stores and used them as Nook screensavers. It's brilliant, not just because it makes reading more intimate, but also because it solidifies the Barnes & Noble brand itself.

And I admit that I love Barnes & Noble and other physical bookstores. An hour spent browsing a bookstore is a day well spent.

It's hard to love Amazon, though. Not the way we love Apple or a bookstore. At best, you respect Amazon for its obsession to detail, for its cheap prices, and for how it achieves the promised arrival dates for its products. You may not love Amazon, but you trust it as a brand. It's sort of like the Post Office. It's hard to love the Post Office, but you never worry much about whether your package will arrive. Although mishaps happen, the Post Office has a great track record.

So for Amazon to launch the Kindle was like the Post Office launching a new e-letter product, a clipboard-sized plastic gadget with a screen on which you can read your letters. You trust Amazon as much as you trust the Post Office, and you absolutely want to read content as soon as it's available. The devices save you trips to the mailbox or the bookstore, and they're excellent adjuncts to your leisure time or business reading. But nobody ever declares themselves a Post Office fanboy and rushes to "unbox" the latest book of stamps.

I should explain that a *fanboy* is a person who's so smitten by a brand that he's often the first in line to buy the brand's newest gadget. A fanboy will often rush home to film the "unboxing" of his new gadget. If you've never heard of *unboxing*, which is the process of unwrapping a new tech gadget while filming it, I encourage you to search this word on YouTube and watch any of the hundreds of thousands of results.

Unboxing is a new voyeuristic phenomenon that's erotic and technical at the same time. It's tech pornography. It's as if we desire total carnal knowledge of our consumer electronics goods. The sheer number of unboxing videos on gadget websites and YouTube is a testament to how obsessed we are.

The fanboys and gadgeteers of our culture are starved for sexy e-readers.

We're a culture that fetishizes technology, and the way people film the unboxing of gadgets is similar to how people ogle lingerie models on the fashion runway every year. How long will it be before we start running new product launches like fashion shows, displaying the new electronic goods on runways with sultry music, paparazzi snapping photos, and the CEO or vice president reduced to someone shilling the product like it's next season's lingerie, a one-handed appreciation of Silicon Valley's newest creation?

There's an economics term for this called *commodity fetishism*. We fetishize a commodity by assuming it's worth more than the sum of its parts. For example, the money we use is worth less than advertised, because it costs mere pennies to make a dollar bill. Beanie Babies at their peak were likewise more expensive than their pure manufacturing value. Consumers perceived that the stuffed animals had higher value, so they rocketed into the status of collectibles, like Cabbage Patch Kids.

The idea of commodity fetishism was created by the wooly-bearded economists of the nineteenth century, making it all the more amazing that they came up with this idea in the age of the horse and buggy. It's still surprisingly relevant now, surfacing as what I would call a techno-commodity fetish that whips people on every year to buy the latest and greatest gadgets.

You can see the techno-commodity fetish in action every time there's a new iPhone. Lines spiral around the blocks near AT&T and Apple stores, and fanboys wait in line for days to be first on their blocks to get the new device. Marketers are savvy to this and play it up. That's why companies like Apple will pre-announce a new iPad: so they can take pre-orders weeks in advance of the new device's availability and drum up demand and, at the same time, very practically let the assembly lines crank out the last of the old devices, using up all the last parts.

By all accounts, this techno-commodity fetish is thriving, judging by the number of gadgets released every year. Barnes & Noble and Apple may have been among the first of Amazon's competitors, but they're not the only ones. In fact, by the start of 2013, there were forty-five eInk-based e-readers for sale, and too many tablets and smartphones to count, all with ebook support. Thanks to booksellers like Amazon and Barnes & Noble, we now enjoy instant on-demand ebooks, which to me is still something fantastic and futuristic, part *The Jetsons* and part *The Diamond Age.*

Bookmark: Book Browsing

The ability to browse for books by their covers and flip through them from front to back before buying is also fading with the ebook revolution. It's sad, but it's a consequence of the way independent bookstores and even major bookstores are faltering and closing their stores.

But browsing through a bookstore is slow—just as slow, in fact, for ebooks as for print books. Whether you're walking down an aisle at your local bookstore or clicking through different categories and subcategories of content on Amazon.com, it's time-consuming.

A better approach than browsing might be something like a Foursquare for bookshelves, so people could become the self-appointed librarians or mayors of a given stack of books and provide recommendations for good books in their section. They could not only function as experts at their local bookstore, but also operate on a regional or national scale. To provide an incentive for good recommendations, an element of competition could be added, with the mayor-librarians defending their turf from would-be challengers.

Imagine checking in once you've read a book on a given topic and developing subject area expertise on something more productive than the microbrews served at the local bar during happy hour. Maybe I'm a tad bookish, but I think an ebook-infused Foursquare would be an interesting idea for a startup. You'd check in every time you read a given ebook and gradually rise in stature within the domain of your expertise in a measurable, uncontestable way. On finishing a book—*ka-ching!*—you'd score points on this online social system. And if I've learned anything from social media, it's that we like to get rewards. They motivate us, especially when our reputation is at stake.

I can especially see this socialization of learning happening now that encyclopedias and other top-down sources of authority have been tossed in the Dumpster in favor of crowd-sourced information like Wikipedia, and as sites like Goodreads and Amazon's own Shelfari democratize book recommendations. But social reading is still relatively new.

Do you trust the recommendations of people online who you've never met? If so, have you discovered a great book through any of these sites? Or better—because we are social, after all—have you met any really interesting people through using these sites? I'd like to hear your story, because let's be honest: you're never going to have an enjoyable chat with Amazon's or Apple's book recommendation software!

http://jasonmerkoski.com/eb/6.html

The Neurobiology of Reading

I've learned a number of things about the nature of reading and e-reading that are essential to understanding where we are in these early stages of the ebook revolution and how much further the revolution needs to go to be truly successful.

First and foremost, e-readers don't hold a melted candle to print books in terms of how crisp and textured their ebooks can be.

When I look at my favorite print books from a tactile perspective, I'm drawn to my childhood Bible, with its thin, translucent onionskin pages like starched Kleenexes. Or my Boy Scout manual with its curiously dated but somehow reassuring 1970s color palette. Or my pulp science-fiction magazines from the 1930s with their brittle, yellowed pages that flake if you turn them too fast and have a texture surprisingly like fiberglass.

The current displays for e-readers are too primitive to adequately convey texture. There's something artistic about eInk, about the almost-random accumulation of tiny titanium dioxide balls in a bath of black ink. But eInk does not produce a warm texture. It's not soft and reassuring like a weathered, slightly scruffy page from an old book. And I can see how poorly an ebook's text mimics the type in a print book. Even seen under a magnifying glass, the type is too pixelated.

No e-reader is able to match the resolution of reality. At best, current eInk readers are able to show 200 dots per inch of resolution, but that's paltry when you consider that even the most mediocre of mass-market books is printed at 300 dots per inch, and photography and art books commonly have two or four times as much texture. If you're on the side of print books, I agree with you on this one. They win on texture, and ebooks still have a long way to go.

There's also a solidity to print books that lends itself well to the gravitas of the ideas expressed within or to the solidness of the story. Moreover, the sense of touch—of pages that are perhaps rough or smooth or crisp or corrugated—gives readers an anchor, continually re-establishes a link between the book and the reading experience, and prevents the mind from wandering while reading. The physicality of a book anchors you to it, unlike the denuded, sterile sensation of sheer plastic or numb glass on an e-reader.

Your brain is aware of this too.

In the brain, reading is as much a sausage factory as the ebook conversion process is. As long as the sausage factory doesn't get choked up, you're able to read each word sequentially. You chunk these words together from your storehouse of understanding about semantic meanings, syntax, and grammatical structures. And as your eyes race ahead to the next word or backtrack a bit to reaffirm what they just read, you have time to think and ponder, to come up with ideas of what the book is about and what you're reading. In other words, to make sense of it.

How does reading work biologically? In a nutshell—and the brain is shaped like a kind of walnutty nutshell, after all—your parietal lobe disengages you from what you were just doing to draw your attention to the words. Your midbrain moves your eye along them, and your thalamus focuses your attention on each letter or word that you're reading. From within the cingulate gyrus, your eyes are directed to each of the words, and then your brain checks to see if the word you're reading is familiar or comprehensible.

Just as a web browser caches parts of a website for faster access later, your brain does the same thing for words. There are caches of visual word-forms for your reference in the part of the brain called area 37 of your occipital-temporal region. Your temporal lobe then translates these symbols into sounds, and the anterior gyrus in the back of your head converts these sounds into your interior monologue, the voice you hear inside your head. Your left temporal lobe and right cerebellum and Broca's area are all brought to bear on making meaning out of this flow of sounds.

It's a complex sausage factory that fits inside your skull, and it moves swiftly, taking no more than 100 milliseconds per word and often less, as long as nothing gets in the way. As long as there are no distractions like strange flickers or ghosts, what you see on an e-reader is just as meaningful to your inner monologue as what you read in a printed book.

Okay, that was all very technical. So let me emphasize the important thing: there's no cognitive difference in reading a sentence in a print book versus a digital book.

However, there's more to a book than the sentences inside it. After a lifetime of habitual reading, your brain is used to considering the whole page of a book in its entirety. Your brain is used to having a dialogue, if you will, with the typographer and page layout artist of the book you're reading. That's why the occasional use of a new font or a drop-cap—or heck, even an *italicized* word—helps you stay focused. It keeps your brain from yawning and switching to something else. With e-readers, though, this dialogue often stutters. The digital page is often bereft of nuance, of any anchor besides a list of monotonously formatted words, like plain black beads on an invisible string.

When you talk to neuroscientists about how the brain works, they'll tell you that a book is meaningless if you don't actively engage with it. That's why poets use unexpected word combinations, or why Friedrich Nietzsche used irony, or why David Foster Wallace used footnotes. These touches disorient you as you read, forcing you to put 10.5 watts of energy into the reading process to actually focus on what you're reading. Why did I say 10.5 watts? It could have been any number, but it was unexpected. It got your attention, and you're more likely to remember this passage now than you would have been if I wrote it in a dry, journalistic way without any memorable facts to catch your attention.

There's something important and touching about the palpable physical presence of a book: it engages the senses. In this way, the act of turning pages helps to anchor information, because we have a visual, geometric sense of where one page is in relation to all the others in a book, a tactile dog-eared map. This is something we lose with e-readers. We're used to processing a 3D world around us in everyday life, and while many e-readers have built-in progress gauges to show you where you are within a book, they're often insufficient.

Such 2D progress gauges require some mental agility to use. They're no better than gas gauges on a car, which show you're halfway through a tank but don't tell you how many miles or gallons you have to go until empty. By comparison, there's no ignoring the handiness of the physical presence of a book as you hold it and the sense of achievement in knowing how far through it you've come.

With ebooks, we also lose the ability to flip back and forth quickly through pages, as we can in a print book. I can flip through perhaps a hundred pages per second in a print book as I look for a given passage, but even on the fastest iPad, I can only see about ten pages per second. Current e-readers are still ten times too slow to match print books in this respect.

So clearly, in some respects, print books are still superior to digital books. Just as we are what we eat, we are what we read. Literally. The act of reading changes the layout of the brain, rewiring it. The more your brain can engage with a book, the better the reading experience becomes, and the more you remember of what you read. And physical sensations—the texture of the paper, the smell of the ink, the raised or recessed letters on the book cover, a peeling price tag on the spine—all help center you in the reading experience and help distinguish one book from the next in your mind's mental map.

That's not to say that e-reading doesn't have advantages, though. And one key advantage is the ability to store and link the books you read.

Some people take the time to meticulously write down and log each book they read, compiling a lifelong list of books that have influenced them. Digital books can not only enable all of us to keep such a list, but also help us do it better.

More than academic curiosity drives people to log the books they've read. In some ways, it's an intimate journal of your mental development. It gives you a ready way to look back on yourself as you were or to retrace ideas to their origins. It may even serve as a memory aid if you're searching for a book you know you once read but subsequently forgot. Also, the act of creating this history helps solidify what you read and anchor it in your mind. It's like you've clicked the Save icon on your word processor and are more likely to recall more of what you read because you saved it to memory.

Ebooks could enable this history automatically for everyone, no effort needed. All it will take is one retailer—say, Barnes & Noble—to add a feature to the Nook that creates a website with a reading history of every Nook book you've read. Every time you buy a new book, it would add to that list, and you could share it with friends and brag about the books you've read.

But I think the biggest boon that digital reading can give us is improved contact between people through better social connections. Reading is often a private experience, and current digital books encourage readers toward even more privacy by allowing them to interact with buttons and joysticks, with toggles and keyboards, instead of directly with other people. It's so much easier to tweet a passage in an ebook we read than to call someone up and talk about it. Digital books are in some ways hastening the lazy, solipsistic narcissism of our culture. We use our gadgets as proxies for other people and genuine human interaction. And yes, I think that's bad.

As a species, we seem to be designed for social interaction, so taking that away leads to problems. For example, research has shown that staying socially engaged keeps a brain vital and fit. A 2001 study published by the American Academy of Neurology found that a healthy social life may cut the risk of Alzheimer's disease by up to 38 percent. Ebooks alone aren't responsible for reducing the quality of our social interactions—we have telephones and chat windows and Facebook feeds to "thank" for this as well—but clearly, e-reading doesn't have to be antisocial.

However, the reading experience can change in the future. It can let you bring your friends or family into the book as you're reading it. Digital books have the promise of giving you the choice, in the moment, of making reading public or private, depending on your mood.

I'll give some examples of these possibilities later in the book. But I want to pause here and agree with print-book lovers out there, because yes, you're right. Digital books aren't quite the same as print books.

Not yet.

Bookmark: Love Letters Preserved between Pages

Feeling nostalgic for print books today, I opened some of mine at random and looked through them. Here's a catalog of interesting bits of stuff I found trapped between the pages:

- A W-2 tax and wage statement from my first job.
- A note from my best friend, torn from a wall calendar dated June 26, 1993, with a note saying, "Jason, come visit!"
- A *Calvin and Hobbes* cartoon my dad mailed me when I was in college.
- Petals from a pear tree I once had outside my apartment in Ohio.
- A record for a Dungeons & Dragons character I once had in junior high (magician, level ten).
- A fax I received on the day my grandfather died.
- Three Chinese coins my mother once gave me.
- A love letter from a former girlfriend.
- A butterfly wing, either carefully preserved or accidentally torn.

The catalog could go on and on. Anyone who knows me will attest to the fact that I'm a collector of useless and sentimental things. My wallet bulges not with cash but with receipts and ticket stubs that I'm too sentimental to toss aside. The result is that my wallet grows larger every month and needs to be held together by too many rubber bands. It won't even fit into my pocket anymore, which kind of defeats the purpose.

I have a habit of stuffing receipts and letters into anything I can find, books not excepted. I leave these trinket collections and stashes of papers behind as bits of myself from former eras. I stuff them into desk drawers and cardboard boxes and wallets and, best of all, books, because there are so many of them in my house. It's as if I'm able to animate the books with my personality, somehow. Maybe I'm a bit pathological in this sense, compared to most people. But books are like waystations in my life, not just in terms of what they taught me, but what they in turn recollect of me, what bits of myself they hold between their pages. There's a capsule history of my life preserved between the pages of my books.

It's a surprising find, and I'd venture to say that everyone with enough books has something similar pressed between the pages of their own books. Such capsule histories comprised of love letters and faded faxes can only be contained by our print books and not by digital ones. Books that I've stuffed with flyers and pamphlets and notes I've jotted on postcards serve as time capsules. They're part of my identity. My identity is not—and hopefully never will be—emotionally aligned with a clean, sleek, soulless plastic device.

In this sense, ebooks are useless.

But now that I'm moving away from print books and toward ebooks, perhaps this is a blessing in disguise. I can reassess whether I even need to store my personal trinkets, my bits of stuff. Maybe it would be best if I simply bought a digital scanner and scanned them. That way, they could coexist with my ebooks on my hard drive and follow me forever, a digital shrine to three Chinese coins and a torn butterfly wing.

But I suspect that something will get lost in translation if these trinkets move from physical to digital. We'll lose the feeling of unexpected discovery. We'll also lose context. Why, for example, was a certain love letter placed inside a specific book? We'll lose something of the ineffable mystery of our lives. But what do you think? I'd like to hear what you have found preserved between the pages of your family's books.

http://jasonmerkoski.com/eb/7.html

Why Books (and Ebooks) Can Never Be Replaced

Why do we read? Besides the nagging voice of my second-grade teacher in my ears, what compels me to read?

In its way, reading is highly ambiguous. What, for example, is Joseph Conrad's *Heart of Darkness* "about"? There are many possible interpretations, but none are definitive. Reading is open-ended, plural, meandering, and imprecise, which can be maddening. So what's the allure of reading?

In part there's the status. Reading is a way to emulate the elite of former ages. The elite members of society—not the commoners—were the readers, and they were the ones with power. Not surprisingly, people wanted to emulate them. And in spite of its inherent ambiguity, reading still has an allure because it works. Reading is still the preeminent mode of consuming information in our culture. It's time-efficient and much faster than conversation. Reading is often solitary and free of distraction—unlike talking or watching a TV show, where a soundtrack and audio effects intentionally manipulate your mood and break any concentration you may have had.

That said, the allure of reading is waning. Books are less of a status symbol now than ever before. Our gadgets themselves are the new status symbols, not what we can do with them. And we seem, as a culture, to crave multifunctional devices. Tablets that surf the web and play games. Smartphones that speak back to you sassily.

If our gadgets can be used for reading ebooks, it's often as an afterthought. You don't see people getting pulled over by the police for reading ebooks on their smartphones. They get caught for text messaging. (Although if I were a state trooper, I think I'd let someone go with a simple warning if I caught him reading a good book while driving.) I think this rise in gadget lust and waning interest in reading presages a decline in basic book literacy.

You might argue that, at the very least, our gadgets are helping us use the internet more successfully. But a 2011 study conducted by the Ethnographic Research in Illinois Academic Libraries Project showed that recent internet-savvy college students performed poorly at basic research skills using Google or other search engines. Reading and book literacy may be necessary prerequisites for learning how to refine information and communicate effectively.

There are convoluted semiotic theories of communication that I won't delve into here, but most such theories agree that information is always encoded, transmitted, and deciphered. For example, an author has an idea that he encodes in English with suitable words. The idea is then printed, and then a reader reads the sentences and tries to decode the meaning of the idea. Errors can be introduced at each step in the process, such as the author encoding the sentence with the wrong word (a misspelling or incorrect usage), or the publisher printing the sentence incorrectly, or perhaps the reader not knowing a given word and therefore being unable to decode the sentence or incorrectly interpreting its overall meaning.

For books, it takes longer to encode an idea than to decode it; in other words, it takes longer to write a sentence than it does to read it. These two sentences, for example, were started at a Chinese restaurant in Albuquerque, improved on while driving to a chile pepper festival near the Mexican border, reassembled a week later during a terrible rainstorm, and edited four months later on an airplane.

Writing is complex, even though the basic units of writing are comparatively simple. We have twenty-six letters, twice as many when you factor in lower and upper cases, plus a handful of common punctuation symbols. That's about eighty different symbols, which doesn't seem like a lot to work with. But consider DNA. Although it only has four basic nucleotides, or four symbols, these encode for all life on this planet, in all its diversity. So writing is complex. With all this complexity, there's a lot of room for error in between the encoding and decoding of this information.

So why do we use books?

Books are good for more than being a barrier against the outside world when you need anonymity and good for more than propping up the occasional table or chair. Books strike a happy balance between price, cost to produce, and efficiency of communication. Pound for pound, few information sources are cheaper than a book. Sources that are cheaper (such as pamphlets) tend not to last as long as a book, so when you amortize the cost of production over time, a book is the clear winner. And because so many books are produced in one print run, the costs tend to be low.

There are more expensive forms of information transmission, but few of us can afford a polymathic private tutor like Pangloss from *Candide* to follow us around. Besides, books offer an improvement on a private tutor because you can read and learn at your own pace, as fast or as slow as you please. Even in a college environment with perhaps twenty students to one teacher, you still can't push a fast-forward button on the teacher to skip the slow parts of a lecture, unless it's a prerecorded lesson. Even then, no visual or audio cues indicate when you get to an interesting part. But you can easily skim through a book to get to the cool parts.

Books are priceless. Without them, we're little more than monkeys who have learned to wear expensive wristwatches and designer sunglasses. We've been elevated into an order above all other animals by books, by language, and by story. Books can give us unattainable orients to yearn for. Books can inspire us toward greatness. Books can give us moral guidance or connect with us in ways that even our friends and families can't. I'm sure you have a few prized books that are almost part of you, part of your identity, books that are worth a tremendous amount to you even though they may be scuffed up and battered or dog-eared and underlined.

Books are essential. And it's important that they don't go away.

But surely we can improve on books, now that we're moving into the digital age. If we could redesign reading, what would it look like?

Books serve many purposes. Sometimes we read for entertainment, and sometimes we read to learn. Sometimes we read for distraction or inspiration or edification or to fight the sheer boredom of a long plane trip. But if I had to distill books down to one core cultural purpose, it would be to teach. Books hold the repository of human knowledge, and then some. Even an innocent romance or mystery encodes social mores, cultural stereotypes, details about a time and a place, and an author's insights into the world. The primary cultural function of a book is to teach, and other functions are simply stylized elaborations and innovations on this core function.

So if books, at their core, are about teaching and learning, experiencing and enjoying, then the best redesign of a book would leverage experience itself. Consider, for example, the experience of walking with your dad through the forest as a kid, as he points out all the trees and their names. As you taste some of the cranberries and blueberries growing on the shrubs, he tells you how they grow and what they're used for. You're likely to learn and remember more from a genuine experience like this than you would from a dry, uninviting text.

Perhaps linear line-by-line reading as we know it will fade to a quaint pastime like butter churning or horseback riding once holographic learning is developed—I'm thinking of a *Star Trek*-ian innovation like the Holodeck.

In that TV show, the Holodeck was a space the size of a large theater populated by holograms—projections of people, places, and objects—with which crew members could interact. If books could be translated into Holodeck-style experiences, you would not read a book by linearly regarding one row of text after another, line by line, page by page, but by directly experiencing it.

Instead of reading about the characters in a romance novel, you would be one of them and interact with the others. The novel would be staged and scripted, and you would be a character in the script wearing period-style clothing. Imagine how history lessons, not to mention global ethics, would be revitalized if you had to participate in a simulation of World War II during school instead of reading about it as a series of dry events and facts. Imagine how many more students might take an interest in algebra or topology if they could experience a Möbius strip by walking on its surface.

That said, Holodeck-style experiential learning isn't on our immediate technological horizon. In the short term, the future of ebooks might look a lot like an evolved version of Apple's own iBook product. With tables, 3D rotating images, embedded multimedia, and multiple typographic options, it certainly engages the eyes and ears. Especially on a device the size of an iPad. I think this is great, although sometimes such multimedia enhancements are distracting enough to take readers away from the main points of the text. Sort of like a PowerPoint presentation with too many bells and whistles and too much clip art.

Experiential reading may well become the next stage in reading's redesign—for certain kinds of books, at least. I think the only time reading will still be preferred in the traditional linear line-by-line style is when it's no longer used for teaching purposes. There are some palaces of the imagination too tenuous to build from celluloid, some stretches of the mind for which no map suffices but the reader's own personality. Some books need to be presented in their original form, and any additional visual or auditory or virtual details would be an imposition, an interpretation. Creating an experiential, Holodeck-style simulation of a book requires one or more people to script the book, and that scripting locks the book into just one interpretation.

Indeed, we see this with movies now. There have been many remakes of classic plays like *Hamlet*, each subtly different, owing to each director's interpretation. Such interpretation isn't just stylistic or related to which actors are cast in the starring roles; sometimes whole scenes are cut. At this point, the play is no longer true to the original. The only way to understand the author's original intent is to read it in the original line-by-line form yourself or to sample multiple directors' takes, hoping that the author's intent corresponds to the average interpretation of all the variants and remakes.

Strange as it sounds, it may be impossible to experience certain kinds of books in any way other than a line-by-line read. Neither Franz Kafka nor Jorge Luis Borges will yield to the virtual, because their books are too much like poetry.

You can't adequately experience James Joyce's *Ulysses* as a movie or a video game. You have to be bludgeoned with it as a book, overwhelmed with the magnificent, inchoate details of Dublin. Paradoxically, the only way to read this book, which takes place in the span of one day, is to read it over a lifetime.

There's no computer graphics studio in Hollywood that can create an ancient monster from an H. P. Lovecraft story, because the monster only lives inside the reader's imagination. To show the face of the lurking horror, the unspeakable dread, would be to tell a different story, and not the one which Lovecraft intended.

Of course, this is just as true for ebooks as for traditional print books. And that's why reading will never be replaced, although it certainly will change.

We can't possibly know with any accuracy what the future of reading will be like. But that won't stop me from guessing. I think ebooks will one day evolve into something like a movie and a video game combined with the authoritative intent of an astute storyteller. I can suggest that it will be wrought and wired so deeply in our brains that the emotions we perceive from the author will be genuine as far as we're concerned.

We'll feel genuine terror or elation, and we'll be transported into another state entirely, half crafted and half real, as any good story should be. After all, the best stories are half true, half how they should have been, and half cloud. I know that doesn't add up, and that's as it should be. The part in the clouds is where you find yourself imagining and wondering what-if thoughts. It's where your temporal and parietal lobes measure out ideas and your brain's limbic system responds with affect and emotion.

The books we read now are laboriously constructed. Their authors are sensitive to rhythm and rhyme, sonance and sibilance, rising and falling action, and intricate symbolism that sometimes takes a team of scholars to decode. We read these books because we understand the codes and conventions. It's like an author carefully wraps something up for us, a present that we subsequently unwrap, and the act of unwrapping is reading itself. We're taught from an early age what the codes are and how to decode them.

Over time, I think a different form of book will eventually emerge, one that's more rooted in the mind itself. Just as authors type or dictate content now, I think the future might hold some sort of high-speed plug that goes into an author's head, some way of taking an author's imagination and converting it directly into a digital format. The same high-speed cables will connect you to the author's original experience. The act of encoding and decoding will become relegated to artistic flourishes, and we'll be able to participate in the more immediate action of the author's own mind and flights of fancy.

The firsthand experience of life itself will come through unmediated by the encoding and decoding that we currently use in books. Words are often the worst culprits in this. They are ornaments that often get in the way of the book. Like shifting, ambivalent snakes, words are capable of so much suggestion and meaning, but they squirm when you try to pin them down.

I anticipate instead that we will be connected mind-to-mind to the lived experiences of an author—such as the experience of nervous anticipation the next time Jeff Bezos stands on a stage to announce a new Kindle, or the terrifying experience of Felix Baumgartner jumping from a balloon in the stratosphere.

Whether they're more inky or phosphorescent in nature, books will follow the human spirit as it endeavors into the unknown. And though books have been relegated behind video games and movies and TV shows for their share of leisure time in America—the average person spends two hours watching TV every day, which is twenty times more than an average person reads—the art of reading will continue, although its form will surely change. Books are being replaced by ebooks, and in turn, ebooks will be replaced by another seemingly science-fictional innovation, but reading in some form is here to stay.

And though the average American only reads seven minutes a day, and that number is dwindling, I'll take it. I'll happily trade an ounce of blood for a moment with a great book.

Bookmark: Indexes

Indexes are a part of the book where ebooks suffer the most. Textbooks and most nonfiction books often have a section at the end where someone intimate with the text has scoured the book for its main subjects and created an index with the page numbers of when those subjects are mentioned. The best indexes are done by hand and are sometimes as lovingly long as a chapter in the book itself.

One of my favorite nonfiction books, the aforementioned *The Road to Xanadu*, is great in part for its index, because it lists such subjects as icebergs and water snakes, opium fumes and alligator holes, green lightning and the horns of the moon. The diverse list of subjects ranges from bacon and beans to demonology and to the palace of Kublai Khan itself, from slime fishes to ice blink, and from Neoplatonism to the noise made by earthquakes.

The current generation of ebooks ignores all the wisdom inside indexes like that one. True, many ebooks have indexes tucked away at the end, but they're rarely integrated with the text of the ebook. And they're not often hyperlinkable, allowing the reader to jump right to a topic. They often just list page numbers instead, which makes no sense, as many e-readers don't even display page numbers! This is a shame, because when you search for a word within an ebook, your Kindle or Nook should be able to use the index to help find exactly what you're looking for. Instead of just looking for whenever a word appears inside the text, which is how e-readers do searches now, they should give first-class treatment to words in the index, rank those results higher.

I think we'll see this improve over time, as innovator-entrepreneurs build out the index feature. Some genres of content lend themselves better to having great indexes—travel guides come to mind. It's hard to mourn the loss of an index—it's sort of like grieving for an Excel spreadsheet—but the index is just as important for ebooks as for print books. And unlike print indexes, digital indexes can benefit from innovation.

After all, it's not hard to imagine a project that crowd-sources the creation of indexes. Such indexes could become collaborative experiences, ways of building community. We see similar bottom-up contributions on Wikipedia or on specialty wiki sites on the web, where fans lovingly edit content to help future fans. Wikis for *Star Trek*, *Doctor Who*, and *Battlestar Galactica* assiduously index each episode of every season's TV show, introducing the places where new characters enter or old ones leave. The same could easily be done for ebooks, once e-reading platforms start to open up and allow collaborative access.

That said, I'm not sure indexes will integrate in a fluid, seamless way with the Holodeck-style experiential books I wrote about earlier in this chapter. Earlier, I mused about which books can (and can't) be made into immersive, experiential ones. As I said, I'm partial to the works of Borges and Coleridge, and I don't think they'll ever translate well into rich multimedia experiences—but what about you? Do you have any ineffable books, inscrutable plays, or downright diabolical short stories that you would feel proud to recommend online as examples of great writing that can never be made into immersive experiences?

http://jasonmerkoski.com/eb/8.html

Igniting Readers at Last!

How will readers engage with one another in the future? How will they engage with authors? And how far away is a future of direct reader-to-reader and reader-to-author engagement?

Engagement takes many forms. For example, my aunt mails my dad a box of mystery books to read every month, books that she's gone through and wants to share with him. My best friend burns audiobooks onto CDs and mails them to me. My girlfriend loaned me her favorite book when we first started dating as something of a test—as a way of gauging my personality by whether I liked the book. The act of sharing a book is a close connection, often as close as a touch and perhaps more intimate.

You can share digital books, but the experience is less warm than when you hand over your favorite paperback. You won't connect with your friend or loved one over the same cover and talk about the same dog-eared pages.

Digital book lending is swift and soulless right now. At least two retailers offer this feature. It's a testament to Barnes & Noble that they were the first to offer this, that they understand the connection one reader has with another through a loved book that's shared between them—because it's Barnes & Noble, after all, that encourages people to get together in their stores and read books on comfy chairs and that hosts book discussion groups that gather like-minded friends of the written word.

The digital experience of book sharing has a long way to go, and it's a bit crippled now. You get a soulless email from Amazon or Barnes & Noble, and then the book magically appears on your device the next time you're within wireless range. Like much in the world of digital books, it's a bit clinical, designed by technologists instead of humanists. But it works, with the benefit that you no longer have to worry about your friend holding on to the book for years and neglecting to return it to you.

When I first started dating my girlfriend and she loaned me her favorite novel, I accidentally ripped the cover off it while reading. That almost ended our relationship right then and there! With ebooks, there's no damage and no worry. The ebook boomerangs back to you after two weeks. That's a lifesaver and a relationship saver.

Ebook sharing demands to be more personalized, though. It should be as personal as sitting with friends in a café or someone's living room. Ebook sharing needs a major innovation that breaks through the glass of Kindles and iPads, shattering the wall between readers. This needs to be something immersive, like perhaps video windows, to provide joint experiences where all the readers are in the same room. This is what we're really looking for when we share a book with a loved one—a connection with that person. We send the book's author out as an emissary and hope to connect over his or her words.

In a way, we need to combine book sharing with book clubs.

The great potential for ebooks is that they can give you the opportunity to share and discuss a given book not just with your nearest neighbors, but with people in distant cities and even distant countries. You'll have an opportunity to talk to them within the book, face to face perhaps, like with the iPad's front-facing camera. You'll have opportunities to become part of social networks that will emerge from the book itself after being inspired by it.

Perhaps Amazon or Apple will acquire a social network of their own and create "channels" within the network, one for each book. This way, there will be a conduit for discussion built right into the reading experience. Perhaps these channels will be moderated by passionate enthusiasts of each book. Members will contribute discussion topics, and perhaps there will even be opportunities for the author herself to jump in and become part of the book circle, available for question-and-answer sessions.

Of course, as with everything socially networked, you're going to eventually see these sites infested by ads and spam, by digital cockroaches you can't quite kill.

Retailers and publishers are currently building out features for the socialization of content through book sharing and book clubs. Retailers benefit from having these features, because they allow content to go more viral and spread through the social networks of the readers. As it is, you can already Facebook and tweet about passages inside digital books. But before long, we'll start to have conversations on the pages with other readers—and perhaps with authors, as well.

I know of at least two publishers that offer the ability for early readers of a book to directly contribute to the editorial process. Readers can comment on which pages work and don't work, and if the author is receptive to their feedback, then the next version of the book can incorporate the readers' suggestions. This is a useful process for shaping an author's manuscript as it moves out of the publisher's editorial process and into the world.

By allowing early adopters to interact with the book, the author (and publisher) benefit from a higher-quality book, targeted better to what the readers want and expect. Likewise, the readers benefit. On the sites I know of, users can often buy the book at a discounted price because they were part of this editorial process. Both authors and readers are given incentives to engage. The only pity is that this process isn't more widespread and isn't yet built into the platforms of any ebook retailer. You have to be a diehard fan and sign up for this service on a publisher's website.

I'm certain this will change over time, though. Especially for nonfiction works, where the author and readers can refine the content of the book to clarify the subject and include topics that the readers really want to learn about. The author and the reader will spend more time collaborating and interacting.

The concept of "authorship" itself, I suspect, will even blur and be diminished as books become shaped by readers themselves. In some ways, for some kinds of content at least, the author is often no more than a privileged reader herself. She can shape the material, but she often relies on conversations with other expert readers to find facts, elaborate on a point, or fill in missing pieces.

We'll start to see books being written and rewritten multiple times—with new endings or new twists or new characters—as the author and the audience engage digitally, something that can't be done effectively with print books. True, you can release a new edition of a print book with an updated appendix and a new chapter perhaps, but in doing so, you often start a new conversation, rather than adding to an existing dialogue.

A digital book will become like a chat room with a community around it. It could come to resemble an online video game, with readers all over the country having an intense online discussion or playing out the plot at the same time, wearing headphones and talking to one another over the internet in real time. Authors will move into the role of directors and orchestrators, and the audience will move into the role of the musicians. The readers will actually write many of the words. The author will choose the venue and shape the narrative in the same way that an Xbox game designer creates the playing field and core graphics that everyone else in the game gets to manipulate and use.

This is an interesting time for books, and there are many ideas in the wild, some of which will actually take root and grow. Many of these seeds will be grown under the care of retailers and under the guidance of publishers, but clearly, many of these seeds will be nurtured by others, such as startups that patch the cracks between what the publishers and retailers offer for ebook reading.

I can see a totally different set of reading features than those we're used to. A lot of these are social features—and let's face it, we are a social species, a tribal people. Whether your tribe is your family, your school, or your work community, or an actual tribe such as the !Kung in the Kalahari, there's something inborn about how social we are. Reading is a solitary act right now, an isolated interaction between one person and a book. The reading experience is at cross-purposes with our inborn impulse for sociability. So what better way to augment the reading experience than to bring social elements into it?

It's no stretch of the imagination to see people camping out on words or paragraphs within a book, carving out domains of expertise. People might do this for the same reason that Edmund Hillary climbed Mount Everest—because of the challenge, because it was there. Someone, for example, can become the expert on the nuances of meaning of this very sentence.

Readers will camp out on a paragraph or sentence in an author's book, staking it out as their turf and defending it when rivals want to squat on that turf with alternative interpretations. I can see people chatting with one another and coming together in conversations that are centered not just around the book, but around a given chapter or section of a book.

Also, as you're reading, you'll see who else is reading, where they're from, and what e-readers they're using. You might decide to reach out to them and chat about this book or this section. The book might prompt you with some starter questions or conversational topics related to it, much like discussion questions in a book club. The chats can be private or public.

If they're public, they get attached to the book in digital format, like transcripts, accessible by other people, as well. In this way, books might continue the Talmudic tradition of commentary, and commentary upon the commentary. It's a tradition started by Jewish scholars between 200 and 500 AD, and it continues to this day. Seen in this way as stories interwoven with commentaries, books will serve as town halls, literate ones where people come together and talk, and their talks will remain for those who come after them, for readers who venture into this conversational thicket months or even years later.

These chats will probably start with text, although you could easily imagine chats happening in a face-to-face way, with video as well as audio. I can even see authors meeting with journalists and interviewers in the actual pages of their books and conducting the interviews within the books, so that the interviews themselves become part of the reading experience. "Meet me in the chapter on the future," I'll say to any journalist, because that's where I'll talk to them, right here on this page. The book can become the home where you'll find the readers, as well as the author.

But even once a book is done being read, the interaction between reader and author doesn't end. Some readers are privileged. They're either authors themselves or cultural influencers. Typically, the reviews they write often appear on the backs of book jackets or in the first few pages of a book as testimonials to would-be readers.

This concept is archaic in the digital space, because by downloading any Kindle book, you're going to be taken past these testimonials. You'll be plunked down right at the prologue or chapter one. The testimonials may well be in the content, but few readers will notice them. The only place for such book reviews will likely be in the pages of legacy stalwarts like the *LA Times* or *The New York Review of Books*, periodicals that are swiftly moving into digital format themselves, making the reviews and testimonials even harder to find as they vie for our attention with pop-up ads and Facebook games animated right there on your screen.

Paradoxically, the arbiters of taste will likely no longer be professional book reviewers but readers themselves, people like you and me. It's a continuation of the trend Amazon started with its own book reviews, in which anyone can contribute a review for a book and the reviews can be as long or as short as you like. The inherent democracy thus provided is a sensible gauge, more sure perhaps and certainly less biased than the most astute of paid reviewers. Amazon has done a remarkable job with this and still has a leg up on Apple and Google and all the others. Even if you've chosen to buy Apple content for your iPad reading pleasure, you'll still often find yourself going to Amazon to read its reviews first.

Interestingly, some Amazon reviews are better than the products themselves—not only can they be entertaining, but they're social commentary too. I'm thinking in particular about the Denon AKDL1 Dedicated Link Cable or the "Three Wolf Moon" T-shirt or Tuscan whole milk, all of which can be found on Amazon.com. I can read these reviews all afternoon long, laughing my ass off. The reviews likely started as reactions to odd products or high prices—the Denon product is a stereo cable that retails for $999, and a gallon of Tuscan milk sells for $45.

Hipsters started writing reviews to mock the products, contriving fictional reasons for why the products are so expensive—with the laughable results that the milk reviews read like those for high-priced wines ("best paired with fresh macadamia nut scones"). And the Denon cable, these reviews suggest, can transmit music from your stereo faster than the speed of light, with the unfortunate side effect of summoning legions of devils into your home.

The "Three Wolf Moon" T-shirt, with its mawkish and unintentionally hilarious design, soared into mock popularity due to hundreds of irreverent hipster product reviews and found itself to be a top-selling item in Amazon's clothing store. In fact, I think Amazon should consider publishing a book of their best and most infamous product reviews!

The digital space has already started transforming the engagement between author and reader, and that process will only continue to accelerate along the lines I described above. How long will it be before we see a book written as a series of comments on an Amazon product review? How long before we see a novel published only on Facebook as a series of posts, a novel that is inherently viral?

Epistolary fiction used to be popular—that is, fiction based on exchanges of letters—but I think we'll start to see more fiction shaped by the forces (and mannerisms) of social networks. This has already been happening in Japan, where the first cell phone novel comprised of text messages was sold in 2003. It became so wildly popular that a franchise of print books, manga, TV shows, and a movie was spun off from it. There are cell phone applications available in South Africa specifically targeted at letting you write—and receive—novels in text-message format.

In cell phone novels, you receive text messages directly from the author. If you've got an unlimited text message plan on your phone, I totally encourage you to try one of these books—just search for "cell phone novel" on the web and look for a book that's interesting! These books are written in a sparse, sublime style. They come at you like text messages from a friend. And yes, they have tension. Intrigue. And suspense. And in some of these, you can write back to the author to ask for clarification or a change in the plot.

For the first time, authors and readers are able to talk directly with one another. Reading has always been a solitary pursuit, and even book clubs have been small affairs. But now book discussions can cross nations' lines. There's no limit to how many readers can cram into a chat room or participate online with Facebook or Twitter. Now, at last, ebooks have ignited the conversation between authors and readers.

If that's not engagement, what is?

Bookmark: Autographs

Personally, I find book autographs amusing. I look at them like calling cards from the late 1800s, which date back to a more demure, genteel time. But that said, I too have autographed books in my collection. And I'm not alone. Many fans and book aficionados collect author's autographs, not just because a signed book is more valuable, but because it solidifies a connection between reader and author. It brings you closer to the work, as close as you can come without being a character in a book.

One day people will talk about print books in a wonderful folkloric way, as if to say, "You know, people once met with the author in person, presented a book to him, and had him sign it with his own hands! In ink!" Sadly, in the ebook world, autographs don't quite make sense. You can sign the back of a Kindle, but that can maybe hold two or three signatures before it runs out of space. More if you use a tablet e-reader, of course. And besides, the autographs will smudge off.

Inventors are even now coming up with complex Rube Goldberg ways of making autographs work digitally, involving complex combinations of Wi-Fi and flash drives and digital cameras and custom software, but there's nothing like print to let you see the nuances of a signature, the quality and personality of an author's penmanship.

True, you could have a feature on an e-reader that lets an author dictate the autograph and say something like, "Dear Mary, you look great today. Thanks for buying this book. Hugs and kisses, Mark Twain," or even something like an embedded video to show you standing with the author, a camera in the back of an iPad perhaps being used to capture you and embed the footage into the book itself. It's a way to take an old metaphor from the past and reclothe it for the future. Rather than trying to get complex systems in place to emulate autographs, I think inventors would be better off creating new features that only work digitally.

That said, I've invented my own system for giving out autographs as part of this book. If you haven't already signed up through any of the links at the end of each chapter, go to the link at the end of this one to get your own autographed book cover. Signing up gets you not only a personalized autograph posted on your Facebook timeline or Twitter feed, but also lots of other unexpected surprises.

Ideally, of course, inventions like this would be built right into the software that runs on Kindles or Nooks. You wouldn't have to click over to a website, because the process would be automatic, built right into your e-reader. And perhaps one day the autograph will even be inserted right into the ebook itself as a permanent part of it. You'll be able to buy special autographed editions, personalized just for you from your favorite authors. Keep an eye out for this over the next few years, as engineering catches up to innovation.

But what do you think? Have you ever tried to get your Kindle or Nook signed? Do you have a collection of autographed books that is preventing you from making the digital transition? Are autographs worthwhile collectibles or afterthoughts best relegated to the digital dustbin? Click this link to get your autograph and join the conversation!

http://jasonmerkoski.com/eb/9.html

Wax Cylinders and Technological Obsolescence

I'm wearing a white smock and white gloves, and the room is utterly silent. I'm guarded by two men, also in white smocks and gloves, who motion for me to sit down. They sit down beside me, one on either side of me. I can't make a move without their permission, but I don't want to make a move. This could be prison. But no, this is exactly where I want to be.

The room has the sparse concrete emptiness of a police interrogation chamber, but it's merely austere. It's got the feeling of a clean room where even a speck of dust or a fallen strand of hair is seen as a holy horror, but it's not a laboratory at Lab126 or anywhere within the curving, clean white halls of Apple in Cupertino. No, this is the Department of Special Collections at the University of California, Santa Barbara, and I came here to see what our future will look like.

Here in the library, a massive digitization project has taken place. More than eight thousand original wax cylinders from the late nineteenth and early twentieth centuries have been digitized at this library. There are recordings here of Presidents William Howard Taft and Teddy Roosevelt, original Sousa marches, and operatic arias sung by the Great Creatore. But the cylinders were made from wax and wood more than a hundred years ago. They're fragile, and they're falling apart fast.

A librarian brings me an original wax cylinder to look at. The cylinder is fitted onto a phonograph, and a scratchy voice comes out of the horn for me to hear. It warbles with static and rises and falls as the cylinder rotates. It's almost like you're listening to the ocean, although you can hear a man's voice in the background, as if he's drowning in the sea of history and shouting distantly for help and recognition.

There are strong parallels between the first e-readers and wax cylinders.

When they came out, wax cylinders were amazing. They were the iPods of the 1890s. They let you listen to music at any time of day, something previously unavailable to anyone (except perhaps those who were wealthy enough to have their own string band commissioned and ready to play at all hours in their mansions or palaces). And yet when we look back on wax cylinders today, they seem primitive.

In the same way, the amazing e-readers that launched the ebook revolution are just as primitive as wax cylinders. For example, when you listen to an old cylinder, you often hear an announcer describing the music that follows. The announcer is practically shouting at the top of his lungs to make himself heard. Recording technology was feeble in the 1890s, so you had to shout for your voice to be recorded. In a similar fashion, the first ebooks had no fonts and no bold or italic styles, and you had to WRITE IN UPPER CASE FOR EMPHASIS!

The original Kindle was bare bones, as well. It basically only displayed black and white text, in just one font and in just six point sizes. The original Sony e-reader was just as bereft from a typographic point of view, and if you were given a choice between a print book and an ebook printed out on paper, you'd be challenged to choose the latter with its monotonous layout and over-simple style. At best, the text could have three different styles—regular, **bold**, or *italic*. Pictures were a bit of a novelty, even for Sony.

This was originally true of print books as well, though. If you're lucky enough to see one of Gutenberg's Bibles in a museum, you'll perhaps be especially impressed by the illustrations, by the flowing capital letters that start every paragraph, richly colored and unbelievably ornate. But they weren't Gutenberg's doing. His Bibles were actually bare-bones text. The illuminated letters, as well as the chapter headers, would have to be added afterward by hand in red ink by the patrons who bought the Bibles. They would hire artists to paint them in, the way we hire tattoo artists to illustrate our own bodies.

New technology always starts out prematurely, but the early adopters adapt it as best as they can and learn how to shout to overcome the technology's shortcomings. And once the technology matures, the old products begin to fade into the past. Wax cylinders are now fragile and falling apart. Every year, hundreds of these recordings get too brittle to play anymore or succumb to "vinegar syndrome," where they turn to liquid. Fewer than 5 percent of the wax cylinders made before 1900 survive.

In a hundred years, you might see print books about as often as you see wax cylinders now, which is to say, rarely. You'd be hard-pressed to find a wax cylinder now, even at an antique store. Print books will fade, and there's nothing fundamentally wrong with this. Likewise, though the horse and buggy was once the most popular form of mechanized transportation, buggies are now relegated to the lawns of old farms as decorations. And one day, if floating hover cars are ever invented, you'll see Ford Mustangs and Toyota trucks abandoned outside those same farms to rust and weather. Technology has a way of shifting, and we're an adaptable species. That's our genius: we do adapt.

The visit to the wax-cylinder library is unsettling to me. Though we now manufacture millions of Kindles and iPads every year, how many of them will survive in a hundred years to play ebooks or MP3 files? I know of companies with vaults where they archive old MP3 players and e-readers. I've been to these vaults, had the glass display cases unlocked for me, and had the opportunity to hold some of the first 1990s MP3 players in my hands.

I've been to a private video-game museum in San Francisco and had the opportunity to play the original game of *Pong* and to play original Atari and Magnavox Odyssey console games. These aren't antique salt shakers or silver spoons in your aunt's curio cabinet. These are tech gadgets that are barely a decade or two old. And yet they're already relics.

In the second *Back to the Future* movie, there's a prescient glimpse at the display window of an antique store, which has an original Apple Macintosh for sale. For moviegoers in 1989, it would have been no more than a joke to see the hot tech gadget of the year as an antique, but there's a bigger question here about durability.

You can still find old Linotype machines that were once used to set type for small-town newspapers, and even after a hundred years, they often work. They had no system software, no brittle silicon parts. Computers will fare less well over time. They rely on electromagnetic memory, which degrades over time, and on a limited number of spare parts available for repairs. For example, the Lunar Orbiter spacecraft of 1966 mapped the surface of the moon to help choose a landing site for the Apollo spacecraft, but once the mapping mission was done, the tape reels with their data were shelved.

Forty years later, scientists realized how useful this data might be for future moon missions, but they found it was nearly impossible to reconstruct the equipment needed to play back those tapes. After years of scavenging through NASA and Jet Propulsion Lab warehouses, they managed to find four rare tape players. Between all four players, they were able to salvage enough parts to get one halfway working. By contacting the retired presidents of former moon-mission subcontractors, they found additional parts and a small trove of repair manuals that one of them had in his garage.

What they lacked, though, was an understanding of the 1960s mind-set—that is, how people in the era of the Lunar Orbiter thought. They lacked the implicit assumptions that 1960s engineers made and that were never recorded in the repair manuals, and they lacked knowledge of how information was coded and decoded onto those tapes. Information science had matured so much over forty years that it was nearly impossible to mentally travel back in time and think the way engineers did in a simpler time.

This particular story has a happy ending. They leased an abandoned McDonald's, set up shop inside, and deciphered the old tape reels like modern-day cuneiform tablets. And we now have stunning digital images of the moon at an unprecedented level of detail—from tapes made in 1966.

The story is bleaker for software, though. At least with hardware, tape reels, and aging wax cylinders, you have something to inspect and work with. It's a lot harder with bits.

A company in Watertown, Massachusetts, called Eastgate Systems seems to be the sole guardian of aging hypertexts from the late 1980s and early 1990s. Before the advent of the internet, these hypertexts were seen as the preeminent form of digital art. They combined text, image, and sound and often did it in a nonlinear way. Reading these hypertexts was a lot like life itself, in that once you made a choice, you were presented with more choices, and you could never go backward. It's a technique that modern video games like *Heavy Rain* and *Nine Hours, Nine Persons, Nine Doors* have rediscovered.

The pinnacle of such hypertexts was a massive project called *Uncle Buddy's Phantom Funhouse,* which had to be loaded from several floppy disks and which contained programming that actually—and deliberately—caused your computer to crash. It was designed to make you aware of the medium with which you were interacting. It would be like the equivalent of seeing this page turn to eInk phosphors and then disappear once you read it, or like having a book whose pages could only turn forward, because the past got destroyed with every page turn.

Sadly, while it's possible to buy *Uncle Buddy's Phantom Funhouse,* it's nearly impossible to read it. You would need an aging Mac computer from around 1990 and a defunct program called HyperCard. Maybe Eastgate Systems will revivify these early hypertexts, which have since been overshadowed by the internet, and make them available on iPads one day. Who knows? But the point is that software does not fare as well as hardware.

Media in our culture fares poorly in general, whether or not it's digital.

Bookmark: Used Books

There's an enormous market now for used books—a recent article in *Publisher's Weekly* puts it in the billions of dollars. Because this market is so big, it allows us as readers to either buy used books ourselves or read used books that have been donated to libraries.

But there's no such thing yet as a used ebook.

All purchases are individual, and DRM makes sure that a book that you purchased can only be read on your own device. Of course, there's a dark undernet for pirated ebooks. As an experiment, I searched the peer-to-peer networks for *The New York Times* bestsellers available this week in ebook format, and I found pirated versions of all of them, contributed by anonymous and technically sophisticated book lovers. And don't doubt that these folks love books, even though they're pirates.

Still, though there's no legally sanctioned and technologically functional used ebook market yet, it's going to happen. I predict it will be started first by a company like Barnes & Noble that allows the resale of its ebooks through other websites—that is, third-party companies who will be able to sell ebooks to you at a discount perhaps, even though they buy the books directly from Barnes & Noble. There's nothing strange about this reseller model. It's used all the time for physical goods. I think the adoption of a reseller model for digital goods will open up a thriving used ebook market.

Perhaps a time period from the original date of a book's publication will have to pass, maybe a year or two, after which the book will be available as a used ebook sale. At that point, the ebook would be available at a reduced price.

Or perhaps the laws of another country will allow used ebook sales, so you'll end up going to an offshore website, like those that run legal online gambling sites, entities headquartered in Bermuda or Turks and Caicos. They'll provide a marketplace for sellers and buyers, and these entities will take a small percent of each sale. They'll be companies small enough to be run by one person, sitting in a beach chair in Bermuda and drinking a mai tai while his servers hum quietly in some nearby warehouse, raking in the cash and storing all the digital files.

I think it would be healthy if we could have a used ebook market. There are even hopeful signs that it may be starting soon; in 2013, Amazon applied for a patent on the sale of used digital goods, including ebooks. Used books—and by extension, used ebooks—would help readers, because more books could be bought at cheaper prices. This lets a reader get more bang from his or her buck. And it would help authors too, because being able to buy used ebooks means the author's ideas and stories are kept in circulation longer. A book, once read, could be liberated from a Kindle or Nook and find another reader.

The drawback to used ebooks—and the reason why so many retailers and publishers are against them—is that they might encourage book piracy.

Piracy is possible for physical books, although to a much lesser extent. If a book is stolen from your house, it could be sold to a used bookstore, which in turn might resell it. But theft of physical books is rare, and it's even rarer when a volume is resold and recognized—although that does happen. In 2010, experts at the Folger Library in Washington, DC, caught a thief who had stolen a first-edition Shakespeare volume from England ten years earlier. The thief had mutilated the book and ripped pages out, but it was still identifiable.

It's much harder to catch a digital book thief. Ebooks lack sophisticated watermarks or other identifying mechanisms, so one digital book looks a lot like another. This means that there's no real way to identify whether a used ebook was resold after being rightfully purchased or illegally copied one or more times.

In fact, because there's no way to forensically differentiate a pirated ebook from a lawfully purchased one, the assumption is that any ebooks not sold by a major retailer must have been pirated. This taints the concept of used ebooks, which is unfortunate. At this point, used ebooks are presumed guilty of being pirated until proven innocent.

I, for one, think we need used ebooks—but what about you? Would you buy a used ebook? Trade one with a friend? Do you wish you could donate some of your own ebooks to a library? Or do you feel ebooks are already priced well and that selling them for less would hurt the livelihoods of authors and publishers alike?

http://jasonmerkoski.com/eb/10.html

Fanning the Flames of Revolution

As retailers like Barnes & Noble and Apple entered the e-reader market, they created increased competition for Amazon. But more importantly, and more positively for Amazon, their success signaled that the ebook market was evolving from mere innovation to full-blown revolution.

And because the ebook revolution is largely technical, the best measure by which to view it is the theory of the diffusion of innovation.

Everett Rogers wrote a book called *The Diffusion of Innovations*, in which he describes the five phases that an innovation goes through as it makes its way through a population. These phases can be used to understand the way consumers approach any new innovation, such as cars or cell phones or computers. Each phase is represented by a group of adopters in the society.

Statistically speaking, the first 2.5 percent of the population to adopt a new technology are called innovators. The next 13.5 percent are early adopters, and the next 34 percent are the early majority. If you add up these three groups of people, you get 50 percent, half the population. The remaining two groups are the late majority, which represents the next 34 percent and, finally, the laggards, with the final 16 percent.

The labels for these five phases in the diffusion of innovation speak for themselves. But to roughly sum it up, the younger and wealthier you are, the more you tend to find yourself on the side of the innovators. The more risk-averse and traditional you are, the more you find yourself with the laggards. Your social status and education level often follow the same progression.

These factors correlate time and time again with every invention that's been studied using the theory of diffusion of innovation. In fact, innovators and early adopters are no longer the only ones who speak about the diffusion of innovation. The theory has become part of the arsenal of tools used by marketers and product managers when they dream up new business ideas or gadgets. Whether you, as a consumer, are aware of this is unimportant to the people who make and market products, because the theory of diffusion of innovation is as real to them as the law of diminishing returns and the 80/20 principle.

Here are a few examples of how the diffusion of innovation works. It took eighty-three years from the time refrigerators were first available in the United States for them to be available to more than half the households, which finally happened in 1940. Flush toilets were invented later than refrigerators, but they took only forty-three years before they reached everyone in the early majority. That accelerating trend has continued with more recent innovations. Home electricity took only twenty-two years before it reached half of all U.S. households. It took nineteen years for radio, fifteen years for TV, and only ten years for the World Wide Web.

As we move further into the future, the diffusion of innovation happens faster. You don't have to take my word for this; any study of the diffusion of innovation shows the same progression of this acceleration of culture throughout history. Perhaps this acceleration is caused by the explosion in innovations than can be reassembled to make still more new innovations. The acceleration can be bewildering if you're unable to keep pace with it. To this day, my grandmother still refuses to use email, and whenever she has anything important to send, she uses a fax.

When are ebooks likely to reach the early majority? Taking a conservative approach and assuming that will take ten years (the same amount of time it took for the internet to reach the early majority), that puts us squarely in 2016, ten years after Sony launched its first e-reader in the United States. I personally think it'll happen faster, because I'm not conservative, and I see the acceleration of adoption rates across innovations. But even if the conservative estimate is right, half the people who read will have an e-reader of some sort by 2016—and possibly a year or so earlier.

By the early majority phase, you're at the sweet spot in the rate of adoption, when the most consistent growth is happening from year to year. My numbers tell me that's where we are now—with the population of readers in the United States, at least. A 2012 report by Simba Information indicates that 24.5 percent of U.S. adults consider themselves ebook readers, and a 2012 Pew Internet study suggests that 33 percent of people in the United States own an e-reader or a tablet. The phase of the most rapid growth is happening now as reader revolutionaries are taking to the streets and subways with their e-readers. As trendsetters and early adopters, they're being seen, and their ebook reading habits are being copied by others. Ebook-only content is helping the diffusion, as well. The ebook revolution is a bloodless revolution that spans all the acres of the imagination, across all of time and space to everywhere your imagination takes you while you read.

Now, no one ever wrote a paean to flush toilets or refrigerators. No one rang a bell in 1950 during the heyday of the television, and no fireworks went off in 2001 when half the population found themselves on the internet. As far as I know, no one's ever written a paean to a cell phone or even an ode to the humble wheel. But the invention of ebooks is different. It's in a rarefied class almost all by itself, because it involves everything aspiring about the human spirit.

The ebook revolution is ultimately about ideas, and in a very real way, we are our ideas. They're the music that flows through our veins, the jolts of electricity that keep one day from blurring into the next. The revolution in reading has a tangible and noticeable effect on us as a population. The Simba Information report also suggests that a title originally only available as an ebook, *Fifty Shades of Grey*, may have been partially responsible for a 7 percent year-over-year shift in ebook reading in the U.S. population.

Modern revolutions are more like microrevolutions. Modern political revolutions follow the same trend toward increased speed as innovations. The revolutions are instigated and completed faster than political revolutions of yore. This is related to the fact that we're online with one another all the time. We're a connected civilization, and this connection is accelerated more by ebooks than by flush toilets or refrigerators. We're revolutionaries with one another because we're linked by ideas, by currents that ripple through our civilization in the books we write and read.

And it's not just that we have more access to books now or that they're available almost anywhere within sixty seconds. The ebook revolution also means that we can take what we've read, and the ideas that have been sparked, and then communicate them at lightning speed to people all over the world—whether through annotations on the ebook or highlights others can see or social network postings on Facebook or Twitter where we can share an interesting passage from an ebook and our comments about it. Wherever we are and whenever we want, we can talk to others around the globe about a book, as if the world is our reading club and the author our best friend.

When I said earlier that the Kindle was one of the two best inventions of the twenty-first century, I meant the *concept* of the Kindle, the concept of a portable e-reader and all the ebooks that can be read on it. I think that other devices since the original Kindle have vastly improved its basic features and added new ones. But they are all rooted in the Kindle. In terms of reading, the iPad owes as much to the Kindle as a smartphone owes to the humble rotary phone.

Like the basic Kindle, current eInk e-readers still have a ways to go as actual gadgets. They may never truly compete with multifunction devices like the iPad or Google's Nexus tablets or even the Kindle Fire released by Amazon. But what we have now is directionally indicative of a future that all book lovers should want to live in, the future of on-demand reading, of having any and every book that's ever been published available to us no matter where we are.

In fact, some books already exist only in electronic format and offer things that print books can't, like the ability to be updated every few days. For a number of years, Kindle's number-one bestseller was a guide to Kindle in ebook-only format by an author who self-published it with Amazon. The author, Stephen Windwalker, updated his book a couple of times a week. Purchasers could redownload the updates at no charge, so they could always have a fresh copy on hand.

This is something that Walt Whitman dreamed about. He revised his greatest collection of poetry, *Leaves of Grass*, nine times in his lifetime, constantly editing and constantly changing the order of poems, adding new ones, and refreshing old ones. He printed this book time and time again at his own expense, which left him broke and barely able to support himself. Even when he was near death, he was only concerned with proofreading his "deathbed" edition. With ebooks, every author can be his or her own Walt Whitman, constantly reclaiming his or her own work by revising it and redistributing it.

Admittedly, today's distribution network for content updates needs to be improved. As a reader, for example, I have no way of knowing whether a new version of a given book is available on my e-reader. We need a distribution mechanism that works like blogs do to push out updates as they become available and then notify us of them. I'm thinking in particular of the way that the iPad and iPhone show a "badge" on the icon of every application to indicate if there's an update or if I have new email or if someone tagged me on Facebook. The same idea could apply to ebook content.

This is the great thing about where we are right now. I'm not speaking as a pie-in-the-sky futurist but as someone who sees technological inevitability. Nimble authors and publishers are able to move fast to take advantage of new ebook features—like animation, interactivity, live chat, location tracking, quizzes, and recipe calculators—that are gradually being added on top of existing features. People in publishing who are smart enough to leap on this accumulation of features are finding themselves with hits, because these features aren't just snazzy eye candy to the digerati, passing memes of the day that get touted in the blogosphere and in some article in *Wired* magazine. These features work because that's how people want to consume information, and they want it all in one convenient package.

But this can only happen if authors and their publishers fully embrace the potential that the ebook revolution presents. I'd like to say that all of them have the potential to grasp it. But as I traveled the country, meeting with publishers, authors, and others to evangelize for Kindle, I began to see that this was not true. Some did not get it at all, and you could sense it in everything they said and did about ebooks. But thankfully, others did. These are the companies that I think will not only survive but thrive in the world after the revolution, while the others will look around in wonder at how quickly their once-mighty empires fell.

Bookmark: Inscriptions

I was a bookish kid. I'd blow my weekly allowance at a bookstore at the local mall every Saturday. That's when the mall's courts and hallways were occupied by people selling secondhand books. They'd set up shop in front of the comic store and the Orange Julius store, not far from the Spaceport Arcade, where you could always hear the eight-bit battle cries of Donkey Kong. Because the books were used, they were often cheap. Hours after entering the mall, I'd emerge into daylight again with a stuffed knapsack, sometimes too heavy for me to carry on my back. I'd drag it along the sidewalk to my mom's car like a refugee fleeing a burning library, ash and sparks in the wind behind me, determined to save as much culture as I could.

Among all these books, I'd often find inscriptions on the first or second page to boys and girls I'd never met from aunts and uncles of all stripes and sizes. The inscriptions were often inked—sometimes with a bold hand, sometimes a frail one—and were usually to commemorate an event. A birthday, an anniversary, or (more darkly) a divorce or bereavement.

Inscriptions are a more personal, lasting kind of autograph. When an author autographs your book, he's often sitting at a table at a book-signing event, working in a rote, mechanical way and hoping to sell enough books by the end of the day to justify his time. It's vaguely mercantile. But inscriptions are the life-soul of families and can often last longer than the families themselves.

For example, there's a collection of inscribed Bibles at Southern Methodist University in Dallas. Stuck in Dallas for a day due to a long layover, I chanced a trip to SMU and was lucky to peruse its carefully preserved Bibles and see the aging ink in them. Some of these Bibles date back to the 1700s and served in their time as birth records. Frontier families in Texas recorded the names and birth dates of their children, generation after generation. The Bibles give a sense of frontier life, of families living far from hospitals and churches, far from the society of anything but cattle and wolves and the hope for a better life in days to come.

Humble inscriptions are important parts of family history. And yet, if and when my own descendants try to reconstruct their family tree, they'll be stumped by digital books. Because you see, digital books can't have inscriptions. I can't give my girlfriend a digital book and write a note on the front page. Digital books are an extinction event for inscriptions. It's like what happened with the dinosaurs. Though they ruled the world, once the extinction event happened 65 million years ago, the dinosaurs died off, and there were no more dinosaur bones in the fossil record. Thus it is with digital books. In the digital fossil record, there won't be any more traces of inscriptions.

It's almost impossible to trace the life history of a digital good. If you download music illegally from the internet, you have no way of knowing who else owned the music file. It could have changed hands a thousand times, ricocheting from Russia to Serbia to France to the United States, from PC to Mac, from one BitTorrent client to another. Despite its travel, the file is still pristine, original, and untrammeled.

Imagine what a digital passport would look like if it could accompany such a file, stamped and counter-stamped with so many international visas! Unless you crack the file open and re-author it, you can't put your mark on a digital good. Could you modify an ebook and inscribe it? Yes. Assuming you can crack the ebook open, you can use any number of tools to add a page to an ebook, but doing so is a hurdle. Actually, it's more like trying to jump over a hurdle while racing uphill in a clown suit and scuba flippers. It's so hard as to make the effort pointless.

I think we've lost something with ebooks in not being able to inscribe them or trace their histories. We've lost a way of learning about ourselves and our families. But then, perhaps this loss is compensated for by the rise in social networks, where one day your great-grandkids will be able to download all your tweets and Facebook posts.

Will there be tools to allow you to inscribe ebooks one day? If so, they'll need to be provided by the retailers and others who control the reading experience of books, and that means Apple, Amazon, and others. You'll need to rely on these retailers staying in business so that the inscriptions you author stay in their clouds. Once these clouds collapse, the inscriptions will likely be lost forever, unless a company one day provides the service of printing ebooks onto paper and binding them as regular old books.

I can imagine a retro company in Portland or Brooklyn doing this, a company run by hipsters in prim mustaches and fedoras, a boutique company that prints ebooks onto paper in the same way that other Brooklyn boutiques publish print magazines as clay tablets.

Each family has its own story, often partly inscribed in the pages of its books. Does your family have a book with an important inscription? A family Bible? Is a chapter of your own history preserved between the brittle pages of an old book? Care to share your story?

http://jasonmerkoski.com/eb/11.html

Innovators and Laggards: The New Face of Publishing

The biggest revolutionaries in the ebook revolution aren't the retailers or authors—or even the publishers. They're the readers, the ones who took a leap of faith and bought the first Kindles or who plunked down six hundred dollars on the first iPads. They're the innovators and early adopters who told their friends and families how good ebooks were, how readable they were, and who bought up ebooks like crazy.

You have to ask yourself, of course, why people bought ebooks in the first place. To be fair, e-readers are sexy, and they're great gadgets. And when innovators get their hands on a great new gadget, there's often a lot of cachet that goes with it, which others adopt. You see the same thing all the time in fashion design and technology—this trickle-down effect of social mores and conventions, fads, trends, and gadgets. But one thing that is different about ebooks is what I call "reader's guilt."

While MP3 players and airplane-friendly DVD players are neat, most of the music or videos we consume are for entertainment purposes. But books are different. You spent years with them in school. You've likely been taught how important they are, and you suspect in a kind of hangdog, guilty way that you should be reading more often than you really do. That's reader's guilt. And that's partly why some users—maybe even you—voraciously buy ebooks. You feel like you ought to. This nagging, guilty feeling may encourage you to give in and buy a Nook.

And let's face it, we have every right to feel guilty for not reading as much as we ought to. According to studies funded by the National Education Association and publisher advocacy groups, the U.S. population is fragmented into two equal groups: half the population reads, and the rest don't read. We're a nation of readers and nonreaders. According to these studies, 33 percent of high school graduates who don't go on to college never read another book for the rest of their lives, and 42 percent of college graduates never read another book for the rest of their lives. Sadly, 80 percent of U.S. families didn't buy or read any books last year.

These numbers scare publishers, of course.

When it comes to ebooks, there are two kinds of publishers: innovators and laggards. During my time as Kindle's technology evangelist, I met plenty of both.

In my travels, I found that, in general, the most innovative, flexible, and successful publishers in the book market were the small and mid-sized ones. They're the ones that have the most to gain, the ones that are willing to take the largest risks. But they're not so small that taking a risk with technology will bankrupt them. I'm thinking in particular about my own publisher, Sourcebooks, a company I first visited years ago when I was managing Amazon's audio and video ebooks.

Sourcebooks was the first publisher to include CDs and DVDs with their print books, bundled as companions to the content. The idea that you could read the poetry of Sylvia Plath or T. S. Eliot and also hear them reciting their own poetry caused a stir when it was first launched ten years ago. Not only was Sourcebooks first to combine text and audio in print, but they also were the first to make the same move with ebooks. I remember working with them to get recordings of poetry slams digitized or videos by Johnny Cash that could be embedded and then seen in an ebook as it was read.

Sourcebooks CEO Dominique Raccah runs the company with as much attention to detail as Jeff Bezos or Steve Jobs. And yet, unlike them, she's nimble enough to adapt quickly and seek inspiration where she least expects it. She's brazen and no-nonsense, the kind of person who'd run a saloon in the Wild West gold rush of ebooks. (Full disclosure: because of their talent for innovation, Sourcebooks was the first publisher that came to mind when it came time for me to pitch this book.)

Based outside Chicago, Sourcebooks has three or four hands in different technology pies, building out enhanced ebooks that seamlessly integrate video and audio with reading and dazzling storytelling, as well as interactive children's books that personalize the reading experience for each child.

Bill, another innovative publisher I know, runs a company that makes travel guides. He totally gets the future of books, even though he seems like a classic old-school publisher. He enunciates clearly, thinks through every word and nuance, and speaks as if he learned rhetoric in college, clearly a dying art. I could sit for hours listening to him in his conference room, which is lined with travel guides to places as remote as Baja and Bali.

I'm not sure what's more exotic, all those travel guides and the worlds they represent, or this voice of grandeur from publishing's past, when publishers were not only eloquent but understood their financial models and kept up to date with technology. That's dizzyingly difficult and complex for most publishers today, considering how overwhelmed they are by all the new pricing models and gadgetry available.

When I talk to Bill about the travel guides of the future and how reading will change, we agree that there will be guidebooks that blur the lines between reading about a place and experiencing it more tangibly, even from another location.

Publishers like Dominique and Bill are looking at creating ebooks that are more like digital applications, because those ebooks can do more than traditional books or even regular ebooks. They see ebooks as interactive and engaging products, with enough narrative or nonfictional glue to bind everything together.

These kinds of ebooks are expensive to make, so you're not likely to see a lot of them, at least initially. Ebooks as applications are sexy, but like the sexiest of creatures, their beauty soon fades. What looks really hot now with all of its techno-trickery will, of necessity, become obsolete in a few years. That's the way of applications. I challenge you to find a computer that will load and run software you bought ten or twenty years ago. Even if you could find the software in CD or downloadable form, the computer's hardware and operating systems will have changed so much in the intervening years that you'd be hard-pressed to get the application running.

This fast pace of innovation is a problem with technology in general. For example, I found a digital tape of some of the earliest writing I did as a kid, from when I'd visit my father's newspaper and write stories on the newspaper mainframe. These stories were backed up onto tape spools, which I have now. But I've searched far and wide, and the only place I can find that has a working reader for this kind of tape is a computer museum in Germany for technology that was still working twenty years ago.

Technology ages. Fast.

The shelf life of an ebook application is only a few years at best. And an Android ebook app has a different kind of code than an Apple ebook app. They're written in different languages, and you have to pay engineers tens of thousands of dollars to port them from one platform to another. Today's hot application becomes yesterday's fossil in the blink of an eye.

Take a look at the fossils in the Burgess Shale Formation, a strip of ancient rock in the Canadian Rockies. These are fossils of creatures that lived 500 million years ago and can't be found anymore on our planet. Some look like winged lobsters or walking accordions with poisonous spines, like manta rays with parrot's beaks or five-eyed worms the size of elephant snouts. They're creatures with body plans so bizarre and befuddling that we'd be terrified if we saw any of them crawling along the sidewalk. But it's through these bizarre bursts of evolution that nature experiments and selects which creatures will survive and move into another era.

Even so, I'm not faulting publishers in these halcyon, gold-rush days of ebooks for innovating and plunking down $50,000 or more to build each interactive ebook application (and that's what they often cost). They're expensive, and everyone from publisher to author tightens their belts on royalties to make these applications happen. But even if publishers don't see immediate profit from ebook apps, the experiences they gain are essential for evolving into the future. In times like this, when the pace of evolution is fast enough to be called a revolution, there are massive changes and die-offs, and the nimble will inherit the earth. The survivors will be those who are agile enough to scamper between the legs of the bigger dinosaurs, avoiding them as they fall.

I've spoken of innovators, but the publishing world has plenty of laggards, as well, including some of the biggest names in the game. While they were once the industry darlings, many of the bigger, more established New York publishers are now the dinosaurs.

Walking into the New York offices of a Big Five publisher is like stepping back in time. Or like stepping onto the set of *Mad Men*. Even when you're out for lunch with the presidents and general managers, you're often in a vermouth-fogged version of the 1950s and '60s, where deals are decided over lunch or sometimes by the quality of your suit tie or class ring.

Success slows some publishers down, making it hard for them to take risks. And just as Amazon is wary of innovating too fast or leaking its secrets, top New York publishers likewise can be very cagey and secretive. I know of one publisher, for example, who paid for a vice president to rent an apartment for a month and lock herself in there, in total secret, with the manuscript of a forthcoming blockbuster book. The vice president had a month to format the manuscript as an ebook by hand. The publisher didn't want to risk giving outside conversion houses the digital manuscript, for fear it might leak.

But all the secrets come out once a year when all the retailers and publishers gather at a trade show called BookExpo America.

April showers bring May flowers, along with rants from publishers at BookExpo America in New York, the nation's largest book event. I'd fly there every May to represent Amazon in talking to publishers about ebooks and innovation. My meetings with publishers outside Amazon's walled garden weren't all pleasant exchanges of ideas and innovation, though. In fact, more often than not, I'd find myself getting yelled at and treated like I was an invading Vandal or Hun.

On one particular day, for example, I was in a basement somewhere in New York City, and a senior vice president of Disney books was screaming at me.

He was at one end of a conference-room table, and it felt like an interrogation. I usually associate Disney with talking animals and spinning teacups and walking brooms, but when you're actually being yelled at by Disney, you see the dark side of the Magic Kingdom. But I don't hold it against them. An hour earlier, I was in the same conference room, but that time, a vice president of HarperCollins was screaming at me. An hour later, another publisher would be yelling at me.

The screams got worse every year, louder and louder. Publishers love to hate Amazon. Even before Kindle, Amazon's relationship with the publishing world was like that of an aging couple. They were forever arguing with one another, but still married after all these years.

It didn't matter what we yelled about in any given year. The next year, it would be something different—but we'd always shake hands and smile when it was all done. The Amazon folks would move on to confrontations with the next publisher, and the vice president of Disney would go on to yell at Apple or Sony. It's a dance we did every year underneath the trade show floor.

On the floor itself, you could get autographs from famous authors, pick up complimentary books or comics, hold the latest e-readers in your hands, and swap business cards with thousands of small publishers and independents on the book-publishing sidelines.

But two stories below the trade show floors—in underground conference rooms laid out like Cuban detention cells—the real wheeling and dealing happened. Everyone's shirts were rumpled with sweat and exertion, and people were pounding their fists on tables. And yet, everyone smiled to themselves, because everyone was getting something from these negotiations.

The same unholy shrieking happens every year at the Frankfurt Book Fair in Germany—the screams are just more guttural. Even at the London Book Fair, now that ebooks have taken off, strained smiles break through the British reserve of once-formal publishers. That's because serious amounts of money are involved every year at these negotiations, and that's true all around the world.

Book readers are mostly oblivious to these backdoor, underground conversations, because the content keeps flowing, and the struggles behind the scenes are just part of business as usual. But they are struggles for everyone in publishing.

You see, most everyone in publishing came into it with an arts background, a degree in writing. These are people who have read Homer and Aeschylus, who can tell the difference between a simile and a metaphor. They can spot a good book when they see one. But nothing in college prepared them for these blood-elevating, stress-inducing fistfights with words.

They came to publishing because of their love of words and their love of language, because of that imaginative faculty we all possess that somehow switches on when we're immersed in a book—when the real world peels away like an ugly scab and we're left with fresh new skin underneath, entranced by this imaginative new world. Maybe that's what kept us going through all those negotiations at trade shows like BookExpo America.

When it was all done, everyone would smile through thin lips and shake hands, and there'd be an invitation to a party at the Flatiron Building, where everyone would get drunk together with Whoopi Goldberg and Spider-Man. All these publishing executives would party with actors and authors and swill manhattans as if Tuesday was the new Friday, but they'd come back to those underground conference rooms the next day, their hangovers pounding in their heads and their fists pounding on the conference-room tables. We reenacted this ritual every year out of misguided self-interest. But if we didn't reenact this, books would have piled up at the publisher's offices in Midtown Manhattan and you'd have had no way to buy your books.

Even though books are moving to digital, events like BookExpo America are as strong as ever. Likewise, the American Concrete Institute still meets once a year at its main trade show, even though concrete is as old as the Roman Empire. Whenever industries are held together by relationships, you'll still find people meeting every year. So we won't see trade shows like BookExpo America fade or move entirely to chat-room windows on computers just because books are going digital. And especially not now, while the ebook revolution is in full swing and the relationships of key players are shifting on a near-daily basis.

There's a triad here between publisher, retailer, and author. Without any of these three, readers wouldn't have any books to read. Authors write books, publishers package and print books, and retailers sell them. You can't, for example, drive to Random House's offices in Midtown Manhattan and ask to buy a copy of *The Lost Symbol.* They're not going to sell it to you, and the security guards will chase you out. Nor can you drive to Dan Brown's mansion and ask him for a copy. He has security guards too. Authors and publishers and retailers are in an intricate dance around one another, orbiting like stars in a triple-star system. It's a complex, convoluted orbit, but this dance is ultimately for readers' benefit.

Publishers need readers. Gutenberg's financiers sent envoys to trade shows in Florence and Paris. They went to promote his new Bibles and drum up pre-orders. We know this because in 1454, an envoy to the court of the Holy Roman Emperor traveled to Frankfurt for its annual fall fair. That year, the buzz was about a man with a new Bible on display that was "absolutely free from error and printed with extreme elegance." According to the same envoy, "Buyers were said to be lined up even before the books were finished."

A year later, cardinals in the Catholic Church were trying to get copies of these remarkable Bibles, but they were sold out to monasteries, churches, and private buyers. So although I have an image of Gutenberg working in his sooty, sauerkraut-smelling workshop, using nothing but ink made by local manufacturers and paper from nearby forests and perhaps even lead and tin from mines right outside his city, commerce was still an outward, centripetal force in his world.

In the early days of the printed book, publishers like Gutenberg served all three functions in the triad. In addition to printing and packaging the book, the publisher would often retail it by taking pre-orders or offering copies for sale to patrons afterward. Publishers were also often authors. Whether they took books from the public domain or commissioned their own or outright stole and retranslated works by other publishers, they functioned as authors for the first hundred years of publishing as we know it.

Over time, the situation grew more complex with the establishment of a class of people who functioned as full-time authors and the establishment of retailers. So although in the early days publishers held all the power, we're in a situation with the printed book now where these three functions are split.

With ebooks, we're seeing the three functions come together again. All the power is being centralized. But the publishers don't have it. The retailers do.

Some innovative publishers like HarperCollins and O'Reilly Media have built retail websites where you can buy and download ebooks directly from the publisher, and Harlequin does a great job with its own retail website. But those are the exceptions.

What has publishers worried most is that retailers like Amazon are getting into the publishing space. Amazon does this in print with its CreateSpace and BookSurge self-publishing businesses, which allow authors who aren't represented by agents or publishers to get their books into print. And it has its own ebook publishing program.

Because publishers go after titles they think will sell well, they usually ignore self-published authors. Publishers have a nose for money as well as talent. Even if they guess wrong occasionally, they're more discerning than not. And their discernment prevents the market from being flooded by books that nobody's likely to read. Otherwise, there'd be nothing to stop everyone from writing their memoirs or books about their cats.

But Amazon turned this practice on its head by encouraging authors who would otherwise be ignored by publishers to join with them, giving the retailer an exclusive on this content. So if one of these self-published books actually does well, Amazon alone has it and can prevent Barnes & Noble or Apple or anyone else from selling that book.

This is especially true for digital books, where Kindle's exclusive file format prevents others from selling the content. Authors are flocking to self-publishing at places like Amazon because they can be assured of greater royalties—often up to 70 percent of the book's list price, for digital anyway. That is pretty good when you consider that for print books, publishers often only pay an author back 10 percent of the book's list price.

As the museum curators of our imaginations, book publishers don't like the undiscerning attitude that retailers are taking, how retailer-publishers like Amazon are just as happy to publish a potential bestseller as they are a book of bad cat poetry. (And believe me, there's a lot of self-published cat poetry. In my opinion, only T. S. Eliot is allowed to write cat poetry.)

Now, retailers have a lot to learn about being book-content curators, but you can see them starting. For example, Amazon has a team that buys rights to popular books and then republishes and repromotes them. And publishers have a lot to learn about retail, but you can see them starting too. Now that they're in charge of their own prices, they have to learn about competitive pricing and how to price content for special times of the year like "dads and grads" sales events.

The book industry is topsy-turvy now, but you can see how retailers might take over publishing, and it's only natural to wonder if retailers will take over the role of authors, as well. You could imagine Apple commissioning authors to write books or hiring in-house talent to create them. You could imagine Barnes & Noble hiring MFA graduates to crank out novels or coming up with a loose affiliate network of independent writers under contract to write content in the way that the popular Dummies series of books does. You could imagine authorship becoming a corporate commodity. And with that, all three functions in the book triad could come together under retailers instead of under publishers, which is where they started in Gutenberg's time.

Regardless of where the book industry ends up, what's clear is that power is shifting. And it's going to shift toward those who understand technology best.

The centripetal force of technology emboldens innovation, increases complexity, and gives readers more options. Regardless of who dominates the triad, readers win. We're in a tremendous time now when content for ebooks is being sought from mainstream and indie publishers, from top-selling and unknown authors, from startups all around the globe, and even from established technology conglomerates like Google.

Millions of texts across hundreds of libraries are being digitized, even tomes from the 1800s with pages often more brittle than pressed violets. Tech companies like Google and the Internet Archive are scanning all of this content so that the future will have these books.

Big Five publishers are moving more or less quickly to accommodate this technological revolution. Some of them are posturing wildly with their arms waving, as if to say, "Yes, I'm part of this!" But in reality, they often just sit on committees and dabble from the sidelines. Mainstream publishers who still take their triple-martini lunches (yes, it still happens) and focus wholly on print books and established relationships between authors and agents and printers are neglecting the new players. The technologists, the software companies, and the entrepreneur-innovators move at a Silicon Valley pace, rather than the nine-to-five life of Manhattan publishers accustomed to taking all of August off as a vacation, as if Manhattan is somehow part of Italy.

I spoke of the Big Five publishers, but perhaps the industry should start talking about the Big Six publishers, because Amazon is in publishing now. In addition to its self-publishing programs, it has a publishing imprint called Encore, which, in its own words, "uses information such as customer reviews on Amazon.com to identify exceptional, overlooked books and authors with more potential than their sales may indicate." It uses crowd-sourced reviews to help make its publishing decisions, rather than relying on the traditional editorial process.

But even if Amazon doesn't serve as a traditional editorial curator with Encore, other companies more than fill the void. And they're not all publishers and retailers.

Ebook innovation is also happening at two other kinds of places. The first is behind closed and double-locked corporate doors, behind walls of security, at tech companies like Apple and Amazon. The second place is on the fringes, right in public view.

To me, the second kind of place is more interesting. That's where passionate inventors come together to show off their latest homegrown e-readers or applications. In my time at Amazon, I found myself more at home with these kinds of people. I'd often fly at the drop of a hat to join one of their conferences. I liked the feeling of frenetic innovation, the fervency of the converted who gather together and create. These were builders. These were my people.

One of the places I'd find myself was at the Internet Archive. Run by dot-com millionaire and former Amazonian Brewster Kahle, the Internet Archive sees itself as a library for all kinds of media—instructional films from the 1950s, public domain ebooks, live concert recordings, even software and old video games from the 1980s. You can download them all for free from the Internet Archive.

The Internet Archive is housed in a beautiful, whitewashed old church by the marina in San Francisco. When you walk inside, you feel something holy. You feel like this is the kind of place that deserves to safeguard our books and music, like that's a holy mission. And maybe it is. There are still varnished wooden pews, even though the chapel has been converted to a massive conference room.

Now that he's made his millions, what Brewster does in life is based on idealism. There's a subtle attitude you can see in someone who does that—a shift, a lightness of being, or something special in his bearing. Call it what you will, but something shows through in someone predisposed to ethical idealism.

Brewster reminds me of an avuncular 1950s propeller-head, someone who would rather be tinkering and building a ham radio in his basement workshop, someone who enjoys the smell of a soldering iron. He was one of those dot-com millionaires who didn't fit his image very well. But man, he loves books! He pays out of his own pocket for a small army of people to scan in old books to digitize them.

He and the other idealists at the Internet Archive are like monks in the Middle Ages, only instead of recopying ancient manuscripts with pen and ink, they use massive server farms that hum underneath the wooden pews. The Internet Archive is like a Google held together by duct tape and idealism.

Brewster organized great conferences, and I'd be the only person attending who represented a major ebook retailer, probably because Apple and the others didn't have time for this. But I did. It was important. I'd be there in the back of the conference, listening to each person as they stood on the stage for a half hour with their PowerPoints, all those university professors and gee-whiz tech wizards and independent entrepreneurs.

You have to understand that all of these people were genuinely interested in books. They were technological revolutionaries, but since they were often millionaires, they were more like revolutionaires. Anonymous though they may be to the eyes of history, these were people who were making the digital reading experience incrementally better. These were people who wanted to make ebooks more hi-fi, who were passionate about such things as style sheets, fonts, and ligatures. These were people who understood that we had to do more than just replicate what print books have given us over the last five hundred years. They knew that for ebooks to work, we'd have to make them better than print books.

When it comes to the soul of the ebook revolution, the smaller, independent ebook entrepreneurs can make contributions that are just as important as those of the technology giants. But the more the revolution marched forward, the more the tech giants began awakening to ebooks.

And eventually, one giant in particular finally awoke from its slumber. A huge, new player made its mark on the book scene—one that was larger than Apple but playing by a different set of rules than anyone else: Google.

Bookmark: Bookstores

There's a used bookstore in Seattle, right in Amazon's shadow, called Couth Buzzard Books. When I talk to the owner, he says he isn't worried about electronic books. "I used to be a teacher," he says, "so as long as children are reading, it's all good." He just wants to ensure that people are reading, which I agree is important. That said, he's going to retire in a few years, so is he worried about the future of print books? He shrugs his shoulders. "It doesn't matter too much."

Maybe he's wiser than I am, but I think print books still matter. A lot.

Though I worked at Kindle for five years, though I own almost every e-reader known to man, though I pioneered the writing of ebooks more than ten years ago, and though I still love my Kindle, ironically I do have problems with digital books.

When I was a student at MIT, I used to love going to the Avenue Victor Hugo Book Shop in Boston. It had cavernous rooms and creaky wooden floorboards and handwritten signs in the aisles directing you to some great reads. Like most independent bookstores, it's shuttered now. In fact, most were shuttered in the 1990s with the advent of mass-market retail concerns like Borders and Barnes & Noble. Consumers got cheaper books and a wider selection of popular books, but they lost access to the more interesting obscure books. They also lost the feeling of connectedness, of being able to talk to patrons and storekeepers who also loved books.

I think this loss sets us back, because sometimes the most interesting books are the ones that are hardest to find. They're the books that Amazon never recommends to me and that even newer sites like Goodreads never get around to mentioning. Sometimes, to find a good book to read, you need to first find a kindred spirit—and that was often the special role filled by people who worked or shopped in independent bookstores.

Some retailers, like Barnes & Noble, still have chairs set

aside in their stores where customers can read and socialize. There are sometimes Tarot card readings, and if you bring your Nook into the store, you can get often get free desserts or coffee from the pastry bar. Fortunately, there are still great spaces where a community can come together around books.

Reading is like an act of bathyspheric descent into the depths of an inky-black ocean. You're alone as you descend into the dark, as you discover strange creatures. On surfacing, it can be a great feeling to share the excitement, to discuss with others all the luminous eels and unexpected fish you discovered in the depths. (And in the best books, you find these unexpected delights inside yourself, not on any page. The best books tell you what you already suspected about yourself but were perhaps too afraid to scrutinize.) Talking about what you've read is a great feeling, whether it's about a fiction book whose characters interest you or a nonfictional account whose ideas intrigue you and that you want to explore with others to make better sense of them.

I don't know whether physical bookstores will disappear in the digital revolution. But for the moment, thankfully, many of them seem able to hang on and maybe even thrive. I will be hoping that they continue to do so. But I also hope they learn from what the online retailers are doing. It's not enough to keep selling books the same way as always. Bookstores will need to adapt and innovate just as much as any tech startup or nimble publisher.

Bookstores are safe havens for intelligent minds and often are free from the tumult of street sounds and outside stress. Bookstores, especially casual independent ones, often have prowling cats, comfy couches, and the feeling that you could spend all day inside, in warmth and comfort. I'm sure you've passed many hours in the aisles of bookstores big and small, and if you're like me, you have some favorites. Care to share any of yours?

http://jasonmerkoski.com/eb/12.html

Our Books Are Moving to the Cloud

I love my library.

It's big enough that it spans the three floors of my house. It's not the fanciest library; it doesn't have recycled tropical hardwood shelves or ornate display cases. There's no bemused librarian sitting there ready to help me find what I'm looking for. In fact, a small warehouse would be more useful and save me from traipsing upstairs and downstairs all the time.

This is why ebooks are so much easier for me. I can flick open my Kindle and search for a word and, within ten seconds, see the universe of content I have and all the books that mention the word I'm searching for. But this is just a scratch on the surface of what a universal digital library could be.

Google comes closest to my ideal for a universal library. With Google, you've got an ever-expanding library right at your fingertips. Moreover, you can upload a list of all your books to Google and recreate your own personal library in Google's cloud.

Everyone in publishing and retail was looking forward with anticipation and anxiety to see what Google would finally do with its own ebook program when it launched in 2010. Would they introduce their own e-reader? Or a tablet? Or something completely new?

Surprisingly, yet staying true to its roots, Google chose to go with a browser-based solution. Google is just dipping its feet in the water, just testing ebooks out. Theirs is a long-range approach. And ultimately, it is well positioned to take on some of the more long-range reading features that are necessary for the evolution of the book, in what I call "Reading 2.0," because Google stores its ebooks in that most ethereal and powerful of places: the cloud.

While flying, I often read the in-flight magazine, which wants to sell me a robotic pooper-scooper, a talking garden gnome, a Wi-Fi-enabled pizza grill, New Age music for my cat, and a machine that will chew my food so I don't have to. It's like a *Lucky* magazine for the business-class traveler with time and money to waste. It will also sell me a CD shelf that can hold five hundred albums, even though the MP3 revolution is already ten years old and every album I own is digitized. Even my Baby Boomer parents have already converted their albums to MP3s!

Why would I buy such a ridiculous shelf and waste space for it somewhere in my house? It's so 1980s, as useless now in the twenty-first century as mullets, Izod shirts, and boom boxes. The same magazine wants to sell me a recycled tropical hardwood bookshelf for my books. But why spend more money than you need to, especially now that our books soar in the clouds, as weightless as a thimbleful of electrons?

How big is an ebook? The question actually doesn't make sense: a digital book is smaller than a fly, smaller than a microbe. It's just an intermittent flicker of zeros and ones on a hard drive somewhere—on a cell phone, perhaps, or on a Kindle. And because a book is digital, you can make as many copies of it as you like, so you can easily back up your digital library in a few minutes.

But if I were to have a fire in my house and lose all of my printed books, I would have to buy them all over again, one at a time. That would be difficult, since some are pretty much irreplaceable at this point. My digital book library is different. I don't have to worry about backing it up, because I know that Amazon or Apple or Barnes & Noble or any other digital bookseller will do it for me. They're in the cloud.

Of course, I still back it up anyway, because I'm fundamentally paranoid about digital content, and you never know, Amazon or Apple or Barnes & Noble may one day go out of business. It's happened before to companies large and small. The great East India Company, once one of the most powerful companies in the world, went defunct in 1874 after almost three hundred years in operation. If you take the long view of history, it's statistically inevitable that Amazon and Apple and other ebook retailers will founder one day. Anyway, with the cheap price of hard drives these days, I can back up my digital library for less than ten dollars.

And if I forget for a week or a month to back up my ebooks, I can still rest easy knowing that they're in the cloud.

There's that word again: *cloud*. What's a cloud? Where is it? Where are your ebooks, and how do you get them back if your device breaks?

I remember when I first discovered the cut-and-paste functions on a computer, when I was a child. All of a sudden, I learned that you could highlight text and cut it out, but it was still there somewhere. It was floating around in the ether, but in a way you couldn't touch unless you knew the magic incantation, which was the paste command. It's a magical concept, this invisible buffer that holds a couple of words or something as big as a whole story and lets you reposition it at will wherever you want.

The cloud, as we know it now, is the same concept but vastly, vastly bigger. And there's not just one. Just like nature with its thunderclouds and puffy white clouds and tornados, Google and Amazon have their own kinds of clouds, and Apple and others do, as well.

Digital clouds are housed in rooms the size of football stadiums that are full of servers, racks of them from the height of your knees to your head, cabinets of computers with screaming fans strained to the breaking point. There are miles and miles of corridors of them in just one building, and often more corridors of them sprawling out into different buildings.

I've been to Amazon's data centers, seen its cloud, walked down its aisles, and had my hair tossed around by the windstorm of exhaust from all these spinning fans. The whirr and hum of hard drives and fans keep the clouds alive. IBM's cloud has servers so hot that they're cooled by water pumped through pipes deep inside these computers to cool off all their cores.

Clouds are in these massive, windowless buildings, often built near rivers so they can be powered by hydroelectric dams. Whole rivers drain and flow to power these clouds. Clouds use more electricity per day than some developing nations in Africa do in more than a year.

These clouds are the warehouses of our digital content. Whether it's Apple's cloud in North Carolina or Amazon's in Virginia, they're always on, ready at the drop of a hat to send you content unimaginably fast. These clouds are connected by massive data pipes of fiber-optic cable to the outside world, to let requests for data in and to pump massive volumes of data out. They're like rivers in their own rights, muddy torrents gushing MP3 files and ebooks.

Clouds are the new libraries. In a digital world, there's no need to put content on shelves. When Amazon sells you an ebook, it's not sitting on a shelf. Digital inventory is totally different from physical inventory. You either have an infinite amount of digital inventory, or you have none at all. As long as a company like Sony has the rights to sell an ebook, it never has to worry about running out of copies. The ebook is in their cloud forever, ready to sell fresh to new customers or to send down to a device if your copy of the book is accidentally deleted.

Our cloud-connected gizmos let us do this amazing dance with content. If your gizmo is about to die, you can always buy a new one and transfer content from the cloud into your new gadget. It's sort of like in the movie *Being John Malkovich*, where people were able to live forever by moving into a new body. Any number of our gizmos can die, but as long as the cloud persists, our culture continues.

When I visited one of Google's data centers, which holds their own ebook cloud, I was amazed. Even though I'd seen these types of buildings before, it was like that scene at the end of the first Indiana Jones movie where the Lost Ark is packed into a crate and taken down distant corridors lined ceiling-high with such crates, a vast warehouse space. But instead of being deathly quiet, the Google data center was humming and throbbing with fast gigabit cables snaking everywhere, a hum of lights and circuitry.

The people who work inside these clouds wear pagers twenty-four hours a day. They get up in the middle of the night when their pager goes off to alert them of an outage or a hard drive that needs to be replaced or a network that needs to be restarted by kicking it a few times. These clouds of data are managed so tightly, with monitors and alarms that go off at the slightest hiccup, that they're more reliable than almost anything else you can imagine. They're definitely more reliable than your library or mine. We have a greater chance of having our houses burglarized and our books stolen than you have of worrying about whether a given cloud gets shut down.

Because of clouds, you can expect to get used to big, empty bookshelves inside people's homes. Personal libraries will move to the web. True, you can put your personal library on any of today's Kindles, but the more you put into your Kindle's memory, the more you will find that searching is slowed down. So retailers will eventually move this search function to the web, where you'll be able to look up words or phrases from any number of books that you have. Some of the more enlightened retailers will show you results for physical as well as digital books, perhaps depending on whether you bought the physical book from them or not.

And you can use the cloud to search inside your books, bringing Google search technology to bear on your own personal library. Assuming, of course, that a given book has already been digitized by Google. And over time, they all will be. Instead of walking your fingers down the spines of all your books to pick one to read, you'll go to a single e-reader sitting on an otherwise empty bookshelf. With just a few taps of your finger against the touch screen, you'll be able to find any of your books from your home or your office, or from the subway or a sunny hammock somewhere in Central America.

Lacking physical proximity to your content will no longer be a barrier to readability. This will be especially helpful if you're a student or you're researching something, looking for the one idea you need like a needle in a haystack of books.

All that remains is for some sort of bridge to be built between what you already own and what's on the cloud, some way of proving to Google that you already own a physical version of a given book. I can imagine an innovator getting into this space and creating a service that lets you send receipts or photographs to Google for books you've already bought.

Once you show proof that you bought a given book, the book would be unlocked on the cloud and yours to read online, without you having to buy it yet again. Because that's the thing: buying a new ebook is only half of what it will take to digitize our personal libraries. The other half is digitizing the existing analog content already in our possession. Whoever licks that problem will make it possible for us to finally become fully digital readers in our lifetime.

I think Google is incredibly intelligent and far-thinking, and eventually they're going to own our personal libraries. They've been working for the last decade on digitizing content, trying to scan all the books from all the world's libraries and place them in their cloud.

I'm personally a big advocate of literacy, and I've got a collector's mentality. Although I know authors who are in an uproar about what Google is doing, I say, "Bring it on!"

When the Sony e-reader was first introduced, it was touted as being able to hold almost a hundred ebooks. The first Kindle could hold a thousand. Subsequent devices increased the amount of storage—but the cloud liberates us completely. I think the cloud is amazing, because it has the promise of storing all the books we've ever owned. Cloud-based companies like Google know this and are building out their clouds to store more and more. You can almost see the iron girders and mechanical struts in the sky, somehow lofting above it all.

This bountiful, ever-expanding cloud seems good, until you realize that it may come with a terrible price. It may mean that we no longer own our digital goods.

Ownership is already a difficult matter with digital possessions, because there's nothing tangible. You can't touch a bit or a byte. But you can at least store a digital copy of an ebook on a drive somewhere by backing it up. In fact, many people advocate doing such backups, even though Amazon and the others have secure copies of your content in their clouds. I think, however, that if publishers and retailers could get their way, you wouldn't even have a digital file. Ebooks would simply be streamed, one page at a time, while you read. There would be no trace of them on your device afterward.

This, after all, is how TV shows have historically worked. You just watch what comes over the airwaves. This is also how Netflix works.

And it's how music services like Spotify and Pandora work. Even the Google Book product works this way. It simply isn't an option to save a local copy of a song or movie. It's in the cloud, and all you're able to do is rent the content. The same may soon be true with ebooks. All you may own are the rights to read a book but not to own a copy of the actual content.

It's a scary thought, with long-ranging implications—and in my opinion, few of them are for the best. We seem to be boomeranging back to the early days of broadcast media, to the time when radio and TV content were streamed over the airwaves and only rarely preserved on audiocassette or videotape.

With this in mind, I think that companies like Google are smart to focus on content first. You can have the best e-reader, but if your content selection is lackluster, you're just going to be a flash in the digital pan. You can be the talk of the town at the Consumer Electronics Show, the yearly trade show for gadgeteers in Las Vegas, but content is a long-tail proposition, and the accumulation of selection takes time. I know this from leading an ebooks team at Amazon. I know how long it takes to digitize all these books.

Though Google got into the game late, you haven't seen the last of them. Because although their strategy makes their results seem low key in the short-term, it positions them perfectly to drive and lead the next phase of reading, what I call Reading 2.0.

Bookmark: Bookshelves

As a kid, I used to enjoy mock living rooms.

The furniture stores of my youth seemed to sprawl on forever, with one mock room following another. Some were decorated in sleek 1980s decor, while others were warmer and more homey. It was an amazing experience to walk through a furniture store and go through one iteration of a room after the next. Endless foyers with endless opportunities for playing board games or watching TV.

I remember the mock living rooms most because they all had bookshelves. I was drawn to the books, of course. Oddly, there seemed to be no rhyme or reason for why some titles were chosen over others for display in the rooms. The books were all hardcovers, as if even in these mock living rooms, it was important to demonstrate wealth and prestige, perhaps as a nod to the "libraries" of the wealthy, special rooms whose walls were ornamented with leather-bound books. In retrospect, the furniture stores likely bought the books by the pound.

Today's furniture stores are more sophisticated and even have cardboard cutouts of computers inside the mock living rooms. But books are still on the bookshelves of these rooms, as if they're waiting for their owners to one day return and read them. Of course, the owners will never return home, since the furniture stores are simply aspirational galleries for homemakers. And yet, I've never once seen an e-reader inside a furniture-store showroom, mock or otherwise.

Old habits die hard. And while books linger on in our cultural consciousness, so will bookshelves.

Is the feeling of warmth we get from a well-stocked library or drawing room genuine, or is it simply something accultured into us? Would we feel just as cozy in front of a fireplace in a room bereft of everything but a Nook set alone on a pedestal? Clearly this feeling comes from our culture. We associate poverty with an empty environment and wealth with a richly appointed one.

But can these perceptions change? Can we collectively become comfortable with simplicity and minimalism? Can nothingness become the new black, as they say in fashion circles?

We've yet to see Swarovski-crystal Kindle cases or cashmere iPad protectors, but as ebooks penetrate into the echelons of the ultra-wealthy, you may see gold-plated styluses and Kindle chargers with prongs made from polished rock from the top of Mount Everest or Mars. Then again, perhaps we will come to appreciate the austerity of style that the cloud brings. If all your ebooks are a download away in the cloud, why display them ostentatiously?

Cicero said that a home without books is a body without a soul. So what does that mean now as we start to relegate our bookshelves to our garages or sell them off at yard sales? Do we not have souls? What does it mean to our spiritual lives if we stop accumulating physical books, these printed volumes that once graced our lives? Are we going to have vast, ornate Edwardian mahogany bookshelves with just a Kindle or Sony e-reader or Apple iPad by itself on one shelf?

As digital goods, books are just pieces of media now, like TV shows and movies and songs and apps, there on a skeuomorphic, digital simulation of a bookshelf on an iPad. "Skeuomorphic" is the word for a design philosophy that Apple, in particular, believes in. It's a philosophy of ornamenting the digital with useless and irrelevant aspects of the physical goods they were copied from. For example, on Apple's iCal product, you can see a leatherette blotter to make the digital calendar seem more like a physical one. Likewise, in Apple's iBooks app, the whorls and burls of wood on a bookshelf have been replaced by a digital texture.

The move toward digital books democratizes fashion and style. It's no longer necessary to buy teak bookshelves, no longer necessary to display your books in a place of pride in your home. Bookshelves are being relegated to that great consignment shop in the sky, where you can also find CD towers, videotape cabinets, decorative typewriter-ribbon canisters, and home darkrooms for processing 35mm film.

Of course, on the flip side, doing away with bookshelves is another nail in the coffin for books, as relics of an elite aristocratic age when we judged one another by what we read. As homes lose their bookshelves, books lose their elite status. In fact, we all lose.

Surprisingly, as a culture, we seem to be okay with this. But what do you think?

http://jasonmerkoski.com/eb/13.html

Google: A Facebook for Books?

When I speak of Reading 2.0, I'm using a metaphor from Silicon Valley. The first release of a software product is Version 1.0, the second is 2.0, and so on.

Reading 1.0 is the experience we're all familiar with: reading a print book from the title page to the author bio at the back. That experience hasn't changed much in thousands of years, not even since Gutenberg's printing press. Reading is still a linear experience, a static experience. Whether you read a clay tablet or a scroll, you read in one direction—from the start to the end of the text.

But now, we're at the threshold of Reading 2.0, a seismic shift, because with this development, reading is no longer linear, no longer static.

Although the idea of nonlinear, weblike reading was discussed as early as 1945 by Vannevar Bush, a computer theorist at MIT, it wasn't until the advent of hypertext and the web in the late 1980s and early 1990s that we saw the first glimmering of this new form, where we could jump around within a book or in a collection of them.

Now, we already understand that ebooks can be fluid. Their content can change seamlessly as they're updated, as the author or the publisher sends down new changes to the text. But more importantly, the texts themselves could be collections of other texts that are constantly changing. Reading could become dynamic.

The reading experience could become more social too. Ebooks allow you to interact with other readers. You can't look at a print book and see who else is reading it and then tap their names to tweet with them about your favorite plotlines or passages. But you could with ebooks.

With all its cross-linked books, Google may be on the verge of making Reading 2.0 possible. They may do it in a number of ways. But my hope and suggestion is that they do so through an idea I have, which I call "the Facebook for Books."

You see, I believe there's ultimately only one book. All books, digital and physical, are part of this one book. No book exists in a vacuum. Even a book of fiction like *The Lost Symbol* mentions outside references. All books are linked in this tangle of intertwining roots, which you can think of as hyperlinks.

In the future, there's going to be just one book, a vast book that includes all the others inside it, which I call the Facebook for Books. You'll be able to start reading from any book and naturally segue into a different one, just by following a link. It could be a bibliographic link or just a link to a book that influenced the author and that's been annotated as such by a reader like you or me. You will be able to link forward or double-back and keep reading. It's social networking, if you will, for books.

For this to work, a critical mass of books will have to be digitized. That's because there's a network effect, a compounding effect. The more content you get, the more cumulative the connections are between books, and the more intertwined and rich the network becomes. A small network will only have a few books and a few connections, while a rich network will be able to link in more. Having more links provides more pieces of the puzzle, more ways of seeing how one book influences or leads to another. And ultimately, readers will have more interesting reading experiences as they follow all these links.

As a reader, you will have a richer appreciation of a book's subject matter and different insights from multiple authors—sometimes contradictory—all just a click away and all of which give you a better appreciation of the whole. This kind of deep linking lets authors have a debate right on the page you're reading, and you get to judge which author or which idea wins. The often-overlooked hyperlink can make this happen. I truly believe that the hyperlink was a twenty-first-century invention that was somehow discovered too early in the twentieth century, an invention we still haven't managed to fully exploit.

Google is in a great place to make this happen. They know search engines, and they can figure out how to continually process the content of all books to make these hyperlinks, so that all the references between books are intact and up to date.

We can already see hyperlinks in the indexes and footnotes of science journals and nonfiction books, as one book bows its respectful head toward another and as one author acknowledges another. But those are just labels. They're not yet working hyperlinks. Such bibliographies and footnotes could form the basis of explicit hyperlinks between books, although you can't click on an entry in an ebook bibliography and go right to the destination. Not yet, at least.

But sometimes these links are more implicit than explicit. As great as William Faulkner is, for example, his writing would be nothing if not for Shakespeare and the King James Bible. In fact, cultural and literary references abound in books. This book, for example, tips its hat to Samuel Taylor Coleridge, *Battlestar Galactica*, Samuel Beckett, Socrates, Neal Stephenson, and so many others.

But as I say, there is only one book, the book of all human culture. It should be possible to seamlessly switch between books, as opportunity permits. For example, in an early chapter I wrote how the product code names from Kindle came from characters in Neal Stephenson's *The Diamond Age*. It should be possible to let readers seamlessly switch over to reading that book, right here, in the middle of this one. That's how web browsing works, after all. If a book is compelling enough—as I hope this one is—then readers will come back after their jaunts and sojourns into other books.

Not only are all books connected, but so also is all culture. It should be possible to create a link from this book into a related *Battlestar Galactica* episode—or at least to a clip from it to show its relationship to the current content you're reading. There should be a hypertextual overlay across all media that lets a consumer flip from book to movie to comic book and back again, as often as the reader pleases, because there is only one book, the book of all human culture. And let me tell you, it's a great book. But it's so long that you'll never finish reading it in your lifetime.

This "one book" is something that we, as readers, would enjoy having, although retailers like Apple and Amazon might object to this—especially if retailers stand to make less money by selling subscriptions to the one book than they would by selling individual books. Publishers might also object to this one book, because they might not want to link their books to one another's.

That's because publishers care about their brands. But let's be honest: a publisher's brand means little these days. Do you want to buy a Random House book, or are you more in the mood for a HarperCollins title? Is that really the question you ask yourself when you're inside a bookstore? No. You look for a book that's interesting, an author you love, or a subject or genre you want to explore, and you often have no idea who the publisher is. The one book I envision would allow this exploration to happen.

Reading 2.0, as I describe it, would give you a conversation with the book and with other readers, as well. For example, let's say you're a *Harry Potter* fan. You've finished all the books but still want to read more. What if you could continue reading about Harry and Voldemort? The feature I'm describing would let you continue reading stories written by others as fan fiction or essays about the *Harry Potter* series and its cultural significance. Linked together as one book, they'd all immediately available, just one page turn away.

Linked together into one vast networked book, just like pages in the World Wide Web, networked books can inform one another. Ideas within books could be related and linked and commented upon. The comments can live on, as can the paths other people take through the books. Maybe you'll find someone who has interesting reading habits, and you'll follow the paths they take in the same way that you might subscribe now to someone's blog feed.

In the same way that YouTube personalities now video-blog about the latest trends, there might be networked book mavens or rock-star book readers you want to follow. Readers can become agents and sleuths and leave inky footsteps behind them as they roam through millions of books. We can follow each other's trails, like veins of copper or maybe gold.

Google will figure out how to monetize all the world's books, in the same way they already make money from you, whether or not you explicitly pay them for anything. You're a gold mine of data to Google, and they already mine your browsing history and chats and every email you care to send or receive using Gmail. They're creating a genome about you and about everyone else. And of course, yes, they're going to use this data for inevitable advertising purposes, which, after all, is how Google makes most of its money.

But if you're willing to overlook the fact that Big Brother won't be a politician but an ad man and that he'll have the face of Google, and you're willing to experiment with the future as an early adopter, then you should take a chance on Google. Because the future of reading belongs to Reading 2.0. And as hard as it may be to see it now, Google seems to be in the best position to build that future.

Not only do they have about the same number of ebooks available as Apple and Amazon and other retailers, but according to a legal affidavit they submitted as part of a recent court case, they have also scanned in twelve million volumes as part of the Google Book project, and they add five thousand more books a day. Google has digitized more of human culture than any other retailer or library. And when it comes to creating a rich network of books, it's the breadth and depth of content that matter.

And you, as a reader, are the one who benefits the most.

You'll get all the books you ever wanted to read in one endless, insatiable buffet. You'll be able to skip and dance from book to book. As it is now, the only textual links in ebooks come from dictionary or Wikipedia overlays, which are a good start but insufficient to encompass the majesty of all human exuberance, art, creation, and imagination. You'll get all books in one reading experience, and if Google is behind this, it might even be free—except, of course, for pesky ads on the bottom of every page.

Bookmark: Book Discovery

How do you find the next book to read?

A better discovery tool than browsing will likely emerge, one that is based a lot on recommendation engines such as those used by Netflix. These sophisticated engines average the kinds of movies that you like to watch, based on your own habits, with those of other Netflix customers and then blend all these viewing habits together to come up with recommendations. This will happen for ebooks too. Hopefully, these recommendation engines will work as well as the savants who run used bookstores who can zero in on exactly the best book for you just by chatting with you and getting to know you.

Some book discovery services like Goodreads and LibraryThing use individually written reviews as the basis of recommendations, but these suffer from one deep shortcoming: well-promoted books get a lot of reviews, and older or under-advertised books get few reviews. Just because a book is old doesn't mean it isn't great. But if a book doesn't get many reviews, the tacit assumption is that the book isn't worth reading—which may not be true. A truly democratic book-recommendation engine could automatically review and rate all books, giving *Fifty Shades of Grey* just as much attention as a hidden gem like *A Voyage to Arcturus*, one of my favorite novels.

Such a book-discovery system should look at the text of a book as the basis for making recommendations. This would democratize the process and let the text speak for itself. It would allow neglected books to shine and put overpromoted books in their place.

With a democratic book-discovery system, readers looking for a particular niche of content could find hidden gems based on an algorithmic analysis of an author's style, gender, and time period; by the kinds of tables or equations inside the book or the places mentioned; by measures of the vocabulary used or ratios of adjectives to nouns; by the percentage of functional words like "of," "to," and "in"; by sentence length and paragraph count; by dependent clauses and dangling participles; by parentheticals and punctuation; by the use of curses, colors, and capitalized words; by the amount of dialogue and number of dashes; by reading-grade level and density of the text; by the cast of characters and the kinds of plotlines; by the use of footnotes and sentence fragments; by alliteration, sibilance, rhyme, and rhetoric; and finally, by quantifying the subjective, emotional experience of reading a book.

To succeed, a company needs to treat this as a deep problem in computer science—a deep and perhaps unsolvable problem, because after all, what you're really doing is teaching a computer how to read!

Democratic book-discovery engines will eventually emerge and become popular, and a whole host of other book-related companies will emerge too. The ebook revolution has spurred an evolutionary explosion of new startups. Some cater to better ebook browsing, some to annotations. Some sell their books on subscriptions, while others serialize them. We're very much alive in the time of rapid evolutionary change, a lot like the way it was in the era of the Burgess Shale Formation.

I could list the names of some of these new ebook companies, but many of them won't last long enough to still be around by the time this book is published. I'm amazed at the number of these ebook companies and their diversity. It's like being a commentator alive during the Cambrian era and pointing out all the creatures scuttling on the sea floor. There's one with fangs! That one has ten eyes! And that one looks like a crawling tongue! They scuttle through the mud too fast to be named, and I can only marvel at the sheer evolutionary diversity, the creative genius, and the deep pockets of their venture-capital angel investors.

Which of these companies will survive? Which, if any, will be here to stay by the end of this revolution? Do you have a favorite or one you're keeping an eye on?

http://jasonmerkoski.com/eb/14.html

Globalization

We were all rebels and outlaws at Amazon. It was gold-rush territory.

I suppose that's only fitting, given Amazon's roots in the Pacific Northwest, the Wild West Northwest. Back in the 1890s, there were towns in the Northwest—they might be lumber towns or mining towns—that would sometimes succeed. There'd be a boom in mining or logging, and people would flood in from all over the country and the world. All of a sudden, instead of just ten dusty prospectors on the streets after the saloon closed for the night, there'd be lawyers and accountants and, yes, prostitutes, all looking to capitalize on rumors they'd heard of untold riches.

Seattle was once the gateway to gold-rush territory, and that still shows as you drive through the old-timey downtown streets. You can see signs on brick buildings that were meant for prospectors a hundred years ago, signs for stores where they could provision themselves with sleeping sacks and hard-tack and pemmican and gold pans before they headed into the Yukon. But now there's a new gold rush in town, the gold rush of ebooks.

This gold rush is heading farther afield than the Yukon. The move is on to make ebooks work for non-Western languages, and it won't be long before you see Chinese and Japanese content look really good on e-readers. Current e-readers were designed for an English-speaking audience, so there's work to do to make the experience great in other languages. That's why Apple and Amazon and all the others are setting up territorial outposts in other countries—in the Middle East and Latin America and Europe and Australia—all across the globe. Each of these companies is intent on establishing itself above the others as the premier player in ebooks and digital devices.

The great game is now on in corporate conference rooms all over Silicon Valley, and anyone with any stake in ebooks and digital content is planning its company's international expansions. Sony was first, another first for them, when they launched their e-reader in England and Germany and other European countries a full year before Amazon. But Amazon started to catch up by launching a dedicated UK device, as well as a universal Kindle that could be used in nearly every country with a 3G network, even on cruise ships out at sea.

The drawback with the universal approach that both Sony and Amazon have taken is that the devices are still English-centric. All the menus and navigation items and user interfaces are in English, which limits the sales of such devices in other languages. Tablet devices are doing much better internationally because they don't have hardware keyboards that need to be customized for each language.

We're going international with ebooks and written content in the same word-for-word way that print books once did.

Printing the Bible bankrupted Gutenberg. His financiers repossessed the equipment, and with the collapse of his workshop and his numerous lawsuits and losses, his workers had no place to go but elsewhere in Europe. Within fifty years of the first printed book, there were printing presses in Germany, the Netherlands, Italy, Poland, Spain, Sweden, France, and England. There were printers flourishing in hundreds of different towns.

As the printers spread throughout Europe in the sixteenth century, they cross-pollinated and learned from one another, much like tech workers in Silicon Valley today. It's not just the ambience of Silicon Valley—with its sunny climate and wineries and badminton courts—that makes it so successful. No, the tech workers go from one company to the other, like bees from one flower to another, rapidly cross-pollinating all of Silicon Valley with everyone's ideas.

I see the same cross-pollination already happening with ebooks.

I see the flux of executives from New York's book-publishing world coming to Seattle and Silicon Valley. I see Apple people coming to Kindle, Kindle people coming to Sony, and Sony people spreading the idea-pollen further afield. It's all incestuous and cross-fertilized, and now that ebooks have been launched, there are no more secrets. You'll be seeing better and better products, and maybe they'll be more humanistic too.

Wealthy readers in the sixteenth century refused on principle to read printed books. They scorned these books because they seemed to lack the humanistic touch of an actual scribe's hand and thought that the printing was too mechanical compared to the natural way a scribe's hand would vary as he wrote out every word. They didn't like the regularity or precision of the printed book and found it less authentic. Though it was cheaper than a handwritten book, the printed book was scorned so much that printers deliberately introduced defects in the fonts and varied them to make the book seem more irregular and less perfectly typeset.

It was a smart innovation, and if a sixteenth-century MRI machine had existed, it would have shown them what we have now learned: that the subtle differences in script and style in a handwritten book are actually better for reading retention. That's because the brain pauses more and the eyes dart around more frequently to disambiguate words, giving the brain more time as it labors over every word to retain its meaning.

And yet, as we know, hand-printed books didn't last long. How many handwritten books do you have in your own library? None? I thought so. And how many traditional print books do you expect to see in the average household a generation from now? None? Exactly.

The shift in taste away from print to digital will mirror the shift from handwritten to print.

The ebook is the second wave in the original print revolution. And this second wave is larger than the original wave that Gutenberg ushered in. It's a wave that has the ability to bring everything together, if it's done well. As experiential products, ebooks are able to contain images and video and audio and games and social network conversations—something that print books can't hope to accomplish.

Not only that, but this second wave in reading can also bring down cultural barriers, like language itself. In the ultimate imagining of ebooks, it will be possible for one book to be rendered into many languages automatically. Likewise, all the comments will be automatically translated into the same language, allowing you and someone in Egypt or Spain to converse as book lovers while reading the same book, without worrying about language barriers.

But we need good language-translation services first.

Globally, there are about 6,900 living languages and at least that many unique ways of seeing the world.

Languages are puzzle boxes. What we do when we speak expresses only a hundredth of what we actually think. We leap from idea to idea with barely a thought, but we only express one idea of the many we conceive. That's what makes conversations so much fun, and books too. They are all about translating, interpreting, discovering, and creating meanings from these puzzle boxes.

The difference is that languages aren't antiques, like dusty, inlaid Chinese boxes with sliding panels. They're made fresh every minute with every new utterance and usage, and they have to be deciphered anew at every sentence. So much is unsaid in a sentence that it has to be puzzled out and reconstructed. The process can often go awry, maybe because the speaker was joking, or using puns or double entendres, or perhaps because the listener misinterpreted what he or she read or heard.

Given all that can go wrong in communicating a sentence, let alone an entire book, it's a wonder that books are translated at all.

And in fact, no translation is perfect. Any skilled translator will perform a deep reading of the book and try to interpret it before rewriting it anew. Inevitably, though—and this is part of the charm of translations—different nuances are brought to bear on the final translation because each translator interprets the book differently. Each translator implicitly refracts the book's meaning through the crystal of his or her own life.

Some translations are more widely read than the originals and have a greater cultural impact. For example, between 1604 and 1611, the original Bible was translated into the English common at the time of

Shakespeare. Named the King James Bible, after the then-current king of England, it abounds with terms we still use—turns of phrase such as "a broken heart" or "a drop in the bucket" or even "bite the dust" —but these, of course, never appeared as such in the original Greek and Latin.

This version of the Bible influenced writers from John Milton to William Faulkner. The text of the King James Bible was carefully crafted word by word by a committee of unpaid but highly devoted scholars who worked on this as a "labor of love." They were men who perhaps never "saw eye to eye" but who "went the extra mile" to phrase the Bible in simple, easygoing speech.

But we're in a digital world now. Since 2009, more books have been self-published every year than published by traditional publishers. In 2011 alone, almost 150,000 new self-published books glutted the marketplace, according to Bowker, a U.S. book trade organization. This is far more books than can conceivably be translated by humans. We shouldn't have to rely on translators, right? Perhaps we're sophisticated enough now that this can be automated and done digitally.

Google, for example, already offers a way to translate a given ebook into the language of your choice. I wanted to see how accurate automatic book-translation could be, so as an experiment, I took a paragraph from this chapter and used Google's translator service to render it into another language (say, Chinese) and then re-translate it back into English. For example, when I translated "Languages are puzzle boxes" into Chinese and back to English, I got "Languages are mystery boxes, old conundrum boxes." To determine the success rate, I took the number of correct words and subtracted it from the total number of words, and then divided by two, since we're translating twice.

I tested a few different languages this way. Scores ranged from 83 percent for German to 65 percent for Japanese and averaged around 75 percent fidelity to the original text. A cynic would argue that this only proves my writing is more German than Japanese. But I would interpret this to mean that, on average, three-quarters of a given book of similar complexity to my own could be translated reasonably well into any language.

What's the threshold for automatic translation? Apple's virtual assistant Siri seems to have a success rate of 86 percent, and people are still complaining, so clearly, we have a few more years to wait before automated ebook translation happens and we're able to achieve the global vision of reading that I described above. Even though they're offered as part of Google's ebook reading experience, automatic translations just aren't good enough yet. But soon, perhaps in the "twinkling of an eye," automatic translations will be good enough to read—but never as good, I think, as those from a skilled human translator.

Of course, the great thing about the future is that we can't predict it. Perhaps an ebook innovator like Google will build a new Tower of Babel, but in reverse, reconstructing it from its rubble across all cultures. It's ironic that the new Tower of Babel might be raised from the squat, windowless concrete building that holds Google's cloud. Google is well poised to do this with its translation software and expertise.

Is it too much of a stretch of the imagination to imagine that Google can rebuild the Tower of Babel from rubble, one captcha at a time? (*Captchas* are those forms you fill out on websites when you have to verify your identity. They usually have a smudged word or two for you to type in.) Most of the captchas you see on the internet are from Google. They're how Google fixes conversion errors in their ebook content. Every time you verify yourself on a website, you're helping Google to decipher one or two words in one of millions of their books.

As a technologist, I'm optimistic that we'll be able to automatically decipher any book fairly well. I think that is stunning, because it opens up a whole new set of authors for me to read! These are authors who aren't commercially important enough for publishers to translate themselves but authors that I would like to read, nonetheless.

Bookmark: Dictionaries

Dictionaries, as we know them, are static snapshots of a culture at an instant in time, defined by a bunch of old men in an ivory tower in Oxfordshire, England. This ivory tower is crumbling, though, and is being replaced by sites like Wordnik and UrbanDictionary.

In my experience, CEOs of companies are often spreadsheet-blooded, boorish, bottom-line businessmen. But company founders are often warm and soulful. They're people like Erin, the founder of Wordnik. She's so sweet that I wonder if she's ever had a negative thought in her life. There's a little red heart on her business card, for heaven's sake.

The former editor of Oxford's dictionaries, Erin started her company to create contextual dictionaries, to scour the web and books and magazines for words and assemble what those words really mean in context, using clues from the content.

Current e-readers—and some enhanced ebooks—often include a dictionary to help you look up words, which is awesome. It's a feature I miss when I'm reading printed books. These days, I often find myself wanting to tap the physical page to select a word and see its meaning.

Having a dictionary built into my e-reader is great, and dictionaries will only get more exciting over time. That's right. You heard me: dictionaries will become downright exciting! You'll be able to bring out the culture's intent as you read with these new internet-enabled dictionaries and encyclopedias, ones that are germane to the book you're reading.

Imagine, for example, that you're reading a Sherlock Holmes mystery from the 1890s. How nice it would be to have a culturally appropriate dictionary to use while you read, a Victorian one that recognizes 1890s slang and brand names. It would help you get more out of the book as you read it and uncover hidden dimensions.

Certain publishers are starting to do this by building limited glossaries into certain enhanced ebooks. These interactive glossaries are seamlessly integrated into the text and pop up with definitions to unfamiliar words or phrases at the tap of a finger. As more and more books get digitized, algorithms will start to mine texts for bits of slang and brand names and other culturally relevant references and automatically assemble them into minimally invasive, dictionary-like resources you can use while you read.

It's a neat reimagining, in which dictionaries are no longer curated by old men in beards like Daniel Webster himself, arranging index cards for decades into one man's vision of what a dictionary should be. Instead, the culture creates its own dictionary. And the more content that's used, the better the dictionary becomes and the more expansive it is.

I can see what people like Erin are doing, and I look ahead a few years to a time when these live online dictionaries replace those embedded in e-readers. I also look forward to a time when such dictionaries perhaps let you see the author's intent as you read, wavering into and out of focus below the iPad's shimmery screen.

But would you even use such dictionaries? Perhaps you think dictionaries already get in the way of your reading experience and you'd rather enjoy the flow of the author's words without interrupting it. Or perhaps you think dictionaries are overkill and we already have enough basic words in our language to use to clearly express ourselves.

A post on xkcd.com presented a plan for the Saturn V rocket but described its components using only the thousand most frequently used words in English. Surprisingly, the description was very readable. There's no word for rocket, so the caption says, "Fire comes out here," and likewise the crew capsule is a "people box." It's sheer brilliance. Just search online for "Saturn 5 top 1000" to see the full plan in all its glory. And while you're at it, let me know what *you* think about the future of dictionaries and words!

http://jasonmerkoski.com/eb/15.html

Language Change: "Whan that Aprille, with hise shoures soote..."

Lingo S changiN fst 2day. tnk u cn kip ^ W it? gr8. NP. f nt, ur n 4 a vvv hrd time :(

Our language is changing, and lexicographers are jumping out of their ivory-tower windows.

The English language is no longer managed by an editorial team in the austere offices of Merriam-Webster, Inc., or the Oxford English Dictionary. Believe me, I know. I've met with their editors at their offices, and the spirit of the English language had fled. The English language is afoot in the world, and she's not going to be penned up again.

New words sometimes used to take decades to trickle into the vocabulary, but now that happens faster than a speeding SMS message. We even have words that aren't, strictly speaking, words. Both n00b and w00t are examples of leetspeak, internet slang that has gone mainstream. One 2012 estimate suggested that 8,500 new words enter the English language every year. Most of these are product names, such as Twitter or iPad.

What will language be like in the future? Will it be some strange hybrid of letters and numbers? Will new words be graced with arpeggios from the extended ASCII character set? Will we find serious works of fiction studded with smiley emoticons? Will the great American novel be written on a teen's smartphone, one text message at a time, and broadcast live on the internet for everyone to read?

Language change is fundamental and unavoidable. That said,

ebooks are accelerating this change. Ebook self-publishing, for example, encourages new words to enter the lexicon faster than ever before. This is because self-published ebooks are usually edited only by the authors and not by traditional editors, a shift in the process that is used at major publishing houses. Unpoliced by vigilant editors, new words from street culture or internet subcultures sometimes slip into self-published ebooks, intrude into the language, and achieve mainstream status.

And this is nothing to worry about.

You see, ebooks will hasten the rapid change in language and aid in its transformation. But let me pause for a moment to explain language change by way of an example.

I was in the hospital recently, visiting a friend recovering from surgery. She was coming out of anesthesia, and I was a little worried. There's always a rare chance with anesthesia that a patient will die in her sleep. My friend had been out for a long time after the procedure and was finally coming round. As she grogged awake, I asked her if she was okay, and what she said sounded like, well, pure gibberish.

I was concerned, thinking she was speaking in tongues or had some brain damage. So I asked her to repeat it, and she did, slower this time. Still gibberish. I was about to fetch one of the nurses, when my friend finally explained that it was the opening of Geoffrey Chaucer's *Canterbury Tales* and repeated it slowly:

> *Whan that Aprille, with hise shoures soote,*
> *The droghte of March hath perced to the roote…*

She had memorized the passage as a kid and had recited it to prove to herself that her memory was still intact after anesthesia.

I didn't recognize that this was Middle English. That surprised me, because I know modern English, and I've even studied Anglo-Saxon English. But I simply couldn't understand what she said. There's a gulf of centuries between Chaucer and us, and to a non-expert speaking one dialect, the other is unintelligible. I very much doubt that Chaucer would be able to understand our use of English, either, although he'd surely be fascinated.

Chaucer was a contemporary of Gutenberg, and since his time,

English has been radically altered and vastly expanded. The Renaissance brought an explosion of Greco-Latinate words into our vocabulary. We can choose whether we want to sound pretentious or smart. We can obfuscate or hide. We can cogitate or think. The older German-tinted English words like "hide" and "think" are still here, but we can use grandiloquent ones too—like the word "grandiloquent" itself. Not only that, but there also has been an explosion of brand-name words, starting in the 1950s. Chaucer would have no idea how to Xerox a PowerPoint—he would accuse you of speaking in tongues. Or *"spekinde tungen."*

And it's not just words that have changed. It's style too.

English has a lilty, singsong quality when spoken. The words go up and down, like a buoy on the waves. You see this in stylized English writing too. But you don't see it in text messages. And you don't see it in business-speak.

I've suffered through countless Amazon deep dives and read reams of business requirement documents that, if stacked sky high, could be a splinter in God's eye. These documents are logically organized, efficient, and detailed and yet devoid of the soul and sparkle of the English language herself. That's ironic, because all of these documents were geared toward the Kindle, toward reinventing reading.

Text messages and the language of corporate documents are just two of many examples of how written English is changing. There's nothing singsong about these styles. They're factual and show how English has been bent. Words are reduced to their bare essentials, and sentences are constructed in business-ese to convey information logically and unambiguously. It's as if we're writing for computers or we ourselves have become mechanized.

Likewise, it's not just the written form of language that's changing. Some would argue that the content of writing is changing too. That we're a culture that regurgitates existing facts and endlessly recirculates them, while our spirit of critical inquiry is devolving. In *The Shallows: What the Internet Is Doing to Our Brains*, Nicholas Carr makes this very point. Easy access to Wikipedia and Google seem to be making it easier for us to find the fast answer, the quick sound-bite. But we're losing the critical skill of inquiry, of diving deep into a subject through source material such as books and forming our own opinions. If a fact isn't at our immediate fingertips or isn't in the top ten results of a Google search, we give up.

By making information universally accessible, ebooks have a great role to play in reinvigorating our critical thinking skills. But nobody has made the text of current ebooks searchable online in a public way. It will be a culture-awakening moment when Google or Amazon or another retailer indexes their ebooks and makes them available on internet search results beyond their own pay-walls. We'll have primary-source knowledge available at our fingertips and undiluted by opinion or Wikipedia editors. Until this happens, our ebooks are still too far away—even if they're only inches beyond our fingertips.

Content is currently buried in ebooks, and typically, only public domain books have their texts fully searchable on the internet. What this means is that books—our greatest repository of knowledge and inspiration—aren't participating in conversations with us online, with the exception of public-domain ebooks that lag by at least ninety years. Social mores have changed. We no longer say "twenty-three skidoo," for example. Much of the searchable ebook content is culturally irrelevant, and that which is relevant is hidden.

By preventing ebook content from showing up in the results of internet searches, we're missing out on some great information. This is most true for nonfiction. Even newspaper and magazine publishers are smart enough to put their content online where it's relevant—but not book publishers. It will take a tidal shift, a sea-change in opinion about ebook pricing models, before this happens. That is sad and short-sighted, in my opinion, because it means that instead of getting expert facts from within books written by professionals, we're getting misinformation and novice opinions when we perform certain kinds of web searches.

For example, "Is creatine powder healthy when exercising?" or "Can I have caffeine when I'm pregnant?" No public-domain 1920s self-help book offers answers to these questions, because words like "creatine" or "caffeine" were not used in these contexts then. But there are hundreds of chat rooms and forums with wildly diverging amateur answers. Publishers would perhaps argue that this information is valuable and that I should just buy the book and read it. True, but how do I even know which book to buy? If ebooks were universally searchable on the web, I'd at least know which one to buy. As it is now, I don't.

I'm not worried, though. This will shift, in time, as surely as language itself. Publishers will relax their objections to making content searchable, and retailers like Amazon and Google will quickly step in to enable this feature. And then we'll be reunited with the words we've been speaking all along.

Bookmark: Dog-Eared Pages

A month before I started working at Amazon, I was in Kansas City, the home of a great old-fashioned, retro-modern printer called Hammerpress. It is part of the vibrant Kansas City arts scene. On the first Friday of every month in summer, all the streets are full of barbecue and ice cream stalls and the art stores and studios open for you to meet with the artists.

When I was there, Hammerpress handed out some bookmarks. These wonderful strips of thick card-stock had been printed using old-time Western fonts and crazy dingbats of the moon and sun and tombstones in black and gold inks. And even though I think bookmarks are as archaic as business cards, I still use them when I read my print books.

Sadly, there's nothing quite so spectacular and well-designed to use when I digitally bookmark my current page. In fact, most of the time, I don't bother bookmarking my digital reading anymore. When I leave a Nook book and continue reading it hours or days later, it knows where I left off, so there's simply no need for bookmarking. Nonetheless, if I wanted to digitally bookmark it, I still could—and sure enough, you see the upper right-hand corner of the screen fold over, dog-earing the page.

There's no such thing as a personalized digital bookmark, though. But then, you could argue that such bookmarks were gimmicky in the print world anyway, just opportunities for salesmen to sell you adjuncts to your reading life that you never needed. Hammerpress will keep doing fine. They make music posters for bands like Yo La Tengo and are not in it for the bookmarks. I don't know any company that is. It's a sensitive soul indeed who will shed a tear for the death of the printed bookmark.

This type of demise has always happened as one technology replaces another. I have to admit that, as a bookish antiquarian and collector, I am sensitive to the passage of these older technologies. But as much as I would love to send a pneumatic tube with a love note inside it to my girlfriend, I know it's impractical and I use email instead.

Still, the humble bookmark could be reinvigorated. Life could be breathed back into it. Instead of appropriating old print metaphors—a dog-eared page—why not reinvent the bookmark? Why not treat it as something at once digital and alive? If the purpose of a bookmark is to remind you of where you are in a given book, then broaden its purpose. Let it act as an agent of your other reminders and to-dos and calendars. Make it an agent of sorts with access to your personal schedule. Give the bookmark a personality, and let it speak. Let it remind you of appointments.

Give it a voice and a personality, and let it suggest when you're reading late at night that you put your Nook away and get some sleep. We speak of dog-eared pages, so why not make the bookmark into a loyal dog of sorts, one that follows you around in your digital life. Let it also bookmark pages in your browser. Let it run off and fetch new information for you, similar to books or websites that you're currently reading. Let your bookmark learn and adapt to your own needs and habits, and you'll find a companion for life that follows you around, that dogs you as you read and travel through wordsome adventures.

But here's the question: would you use such a digital bookmark? Would you trust it to find good reads for you? And do you even want your e-reader conspiring to make decisions about you behind your back?

http://jasonmerkoski.com/eb/16.html

Education: Print or Digital?

The ebook revolution is ultimately about culture change. It's about the impact of digital books on our civilization and what ebooks mean for you and for future generations. Are digital books an improvement, an advancement that will change how we read and absorb information and ideas? Or were we better off with the printed form, the dusty books we've held and loved throughout our lives?

The answer, of course, is yes to both questions.

Digital books are the closest we've ever come to Plato's ideal world. They're immaculate and reborn fresh every time they're downloaded to a new device, like Cylons in *Battlestar Galactica*. Because of this, digital books are a great fit for schools. Ebooks never get lost or defaced. Schools no longer need to replace books if they're the casualties in a food fight or if the proverbial dog ate them along with a child's homework. It's going to be a lot harder to blame a dog for eating your ebook or hard drive.

Children are highly adaptable by nature, and with the exception of the almost blind, I've never met a child of reading age who couldn't get into an ebook. As adults we may prefer to cling like Socrates to the old way. But trust me, we can all "get into" an ebook. There's no barrier in the brain to reading once you're engaged with a book. And if you say there is, if you genuinely feel that you can't get into an ebook, then it's probably not written well. If you give yourself a chance, you can adapt to the ebook experience. Children who are brought into ebooks now have the golden opportunity to start fresh without any preconceptions.

Now, I mention Socrates because he's relevant to this discussion about the barrier between new and old ways of reading. If you think there's a divide now about reading print books versus digital books, consider that in Socrates's time, there was an argument about the value of reading itself.

Socrates was born into an oral culture, and his teachers taught him through dialogues, which were texts that they had memorized. Socrates learned early on to challenge and question those texts. He was the last philosopher of Greece's oral culture.

His student Plato was brought up in the oral culture but had learned to read. Ironically, it's only through Plato that we know about Socrates, because Socrates didn't believe in writing. He never learned it and never wanted to commit himself to paper. Plato disobeyed his teacher and secretly wrote down Socrates's teachings.

In his day, Socrates was one of the most respected (and notorious!) teachers of them all, which is why I think his words are appropriate here. He lived in a time of incredible change, when the Greek alphabet itself was first developed. (That's an amazing innovation in its own right, right up there with hyperlinks as one of civilization's most mysterious and unexpected inventions.)

While writing existed before the Greek alphabet was invented, there were no vowels. Greek writing, though, was invented with a one-to-one correspondence between letters in the alphabet and sounds that people would pronounce. Greek was simplicity itself. It was immensely efficient in a way that any Amazon engineer would appreciate. And yet Socrates still railed against it! (Although, keep in mind that Socrates was also skeptical of pockets and preferred, like many others in ancient Greece, to keep his money in his mouth. This is true. He would often hold his money in his mouth while walking around and take it out to talk.)

The arguments Socrates had against reading are relevant and deep, and you should get to know them. He argued that by reading, we were too lazy with what we learned. We would say that we had learned something because we read it, but we hadn't actually pondered or questioned it the way he would, the way someone in an oral culture would when memorizing a text, by constantly listening to it and internalizing it and gradually challenging or accepting it. Socrates felt that this act of questioning was of supreme importance to personal growth.

Although I'm an ebook evangelist, in many ways I agree with Socrates, because there's more to school than memorizing facts. I'm of the opinion that a dialogue process is important with any book, that you need to wrestle with the book (or ebook) and what the author is trying to say.

The same arguments Socrates made about reading itself apply to the digital. He'd be out in the streets right now, complaining about the lack of critical skills in children and their inability to think critically about what they read on the web. You might want to read what Socrates said in the *Phaedrus* and come to your own conclusions about whether we should read and how. If after that you still believe in reading, then there's no barrier to digital reading.

If you look at the true importance of what books mean to our culture—and I mean human culture, all culture—then books, in many ways, are what separate us from other animals. Books educate. They convey culture. With a book you can set down all your wisdom and accumulated learning for posterity, and others can read your book long after you've passed on and still learn from you. This is how cultures grow—exponentially fast.

You can't get this without writing. It's just that simple. There's a limit to what you can teach person to person through conversation alone and to what the listener can remember and build on from their recollections. And true, you can still say a lot in an oral culture such as preliterate Greece, the same culture that gave birth to Homer and his incredible blind recitations, inspired poetry of the Iron Age.

Homer's poems were entirely oral, and like him, a diminishing number of preliterate poets still recite heroic oral stories and thus convey the core concepts that define their cultures—concepts like nobility, fighting for what's right, and truth and justice. But it's much harder to educate someone about the art of metallurgy or statecraft through an epic poem. It's nearly impossible to teach medicine or any other science without having a text, something large enough and capable enough to hold the sheer volume of details.

We're unique as a species, we humans, because we created books as educational tools to augment the little that we can convey orally from person to person. There's as much of a distance between our Stone Age ancestors and the preliterate Greeks as there is between the Greeks of Homer's age and the literate billions who now inhabit the earth.

Language is responsible for an explosion of culture and vibrancy and human richness, but it was made exponentially richer by writing, whether in the form of books or scrolls or cuneiform tablets. We're not born with all of our culture's teachings inside our heads, the way animals are born, the way animals know instinctively what to eat or what the shadows of their predators look like. Animals rely on instinct, but we rely on being educated, on stories and tales told by mothers to their children or grandchildren. We put these stories down in books so they can educate any number of generations who follow, and we rely on these stories.

We're born with enormous brains, but we're born without instincts for self-preservation. Baby ponies and lambs can start walking and eating a few hours after they're born, but we take years to do the same. Large as they are, our eggshell-fragile skulls are too small when we're born to hold the wealth and weight of our culture, and it's not passed down through the generations by instinct alone. We rely on culture to teach us even the most basic things, like how to groom ourselves or bathe or eat and drink. And likewise more sophisticated skills, like hunting or agriculture. These cultural inventions are learned and taught, in turn, to the next generation through books.

We've come far in our culture, to the point that we now have digital books and can pick one from millions on a whim and begin reading it in a minute. The pace of technological change—though thrilling—is often confusing. And you can feel like you're never quite caught up. You can subscribe to a hundred news feeds, if you know how to do that, and you still won't be caught up, because the pace of technological change outpaces even specialists in the field.

It's no wonder that a lot of the people I talk to are confused by ebooks. They don't know which way to turn, which page to turn, which e-reader to use, or why they should even use them. And I totally empathize about how confusing technology can be. But technology is just a tool, like hammers and nails, although fussier, more prone to crashing, and more in need of firmware updates and special USB cables.

Once you get your head behind the ebook revolution, once you untangle yourself from all the different power cords and USB cords and actually start reading an ebook, I think you'll realize as I did how useful these books are for culture, for reading. Ebooks, more than print books, offer an immediacy of meaning. After all, a dictionary is built into most e-readers, so the definition of an unfamiliar word is usually just one click away.

If this alone isn't an educational improvement, then consider communal annotations and how they help readers to understand a digital textbook better. Each reader can make their own annotations to the same digital book, and all annotations across multiple readers can be added together. Some e-readers, like Amazon's, show you the number of times that a given passage has been annotated. There's often a wisdom to crowds, and in many cases, the most frequently annotated lines in a book are the most salient, the most useful for learning that chapter's point.

This, though, is the paradox of ebooks: if you accept that children should read and that ebooks can teach as much as a print book, why didn't we digitize textbooks first? Because we didn't. Instead, we digitized fiction, sci-fi, romances, *The New York Times* bestsellers, and yes, pornography. Stuff we knew we could sell. But it's content that hasn't reached children in a significant way.

This is the central paradox of our ebook revolution: digital content won't really succeed until it's part of our culture from a very early age, and I mean from first grade onward, from the time children start reading. E-readers need to be flexible and sophisticated enough in their features to allow that. Right now, they're just not adequate.

There are some neat experiments—as I write this, for example, I have friends in the publishing industry who have quit their jobs in Manhattan and gone to work in Silicon Valley for a company that builds e-readers for students. These devices have two folding screens, side by side like pages in a book, that allow you to write and scribble and draw and download and read books.

Tech experiments like this are what we need to really make education work digitally. And until we do that, ebooks will be something that's bolted on to our culture. Ebooks won't really be part of our culture until we're raised with them, until we're digital natives who stare with newborn eyes at these phosphorescent eInk displays.

Of course, a part of me yearns for good old-fashioned print books. And if I ever had a child, I can see how difficult it would be for me to choose whether to let the child read ebooks or use a computer or even have a smartphone. I'm sensitive to these issues, and a lot of parents I talk to also are worried that their kids will be distracted from reading by videos or social networking apps on an iPad or screeching monkeys in a game built into an ebook.

Teachers are worried too.

Professors are bemoaning the loss of critical thinking skills in today's students and the loss of active reading skills. When we passively consume content, lazily let our brains stop doing the hard work of reading, and turn instead to the distractions of tweets and games, we're changing our brains. We are what we eat, and the same is true of our digital diet. We are the media we consume, distractions and all. In the Stone Age, our ancestors listened to birdsong and bee hum, and that was media enough for their minds. Then we developed song and story. But now we're no longer content with the oral tradition, as Socrates was, nor are we content with reading and writing. We want distractions. And we want digital distractions most of all, because they're convenient, downloadable to our devices in under sixty seconds.

In fact, our habits for digital distractions and passive content consumption are putting us in danger of becoming a new species.

I'm not saying that we're going to become robotic Cylons. But we are in danger of becoming a species whose brains are wired totally differently than the humans who came before us. A species that can't reason critically, can't engage in active imagination, and can't read into a mystery and figure out who murdered the butler before the novel ends. With the increasing interconnectedness that our devices afford us, this new species is likely to be much more social, like hyperactive orangutans on Facebook. I can't say what this new rewired species is ultimately capable of. Socrates himself couldn't say what the future of reading and writing would hold. He just rejected it wholesale.

We don't need to reject digital culture altogether. We just need to be careful. Stick to dedicated experiences and be wary of digital distractions. Set a time limit for the amount of time you or your children use in consuming media. Resist the impulse to tweet something every ten minutes. (It takes your brain at least twenty minutes to focus itself again after a distraction.)

It's easy to say that digital content is *not* a good thing, especially for a developing child. I myself once believed this. But now I think this is overly simplistic. If you're objecting to the new merely because it's new, you become an old stick-in-the-mud like Socrates.

Just as there was a gap between oral and written cultures in the generation between Socrates and Plato, there's a gap now between analog and digital cultures. All of us sit squarely between both analog and digital cultures. We were raised on TV and print books, but we also had computers and the internet. We see the allure of the digital culture but still remember what it was like to use public pay phones. We're hybrids. Neither fully analog nor fully digital, we're able to pause on the brink of this digital gap and look fondly back to phonebooks and pennies and other ephemera of an analog era. But now we turn toward the digital future, toward credit cards instead of cash and ebooks instead of print. The digital culture is upon us, and our children will be the heirs to a fully digital culture.

What will the future of education hold?

It's more than simply taking old print metaphors and making them digital. The future of education isn't about virtual blackboards or playing learning games as a kind of digital recess. I actually think we're going to see more social elements in education. And let's just accept the inevitable: social networks like Facebook and Twitter will be available for children at some point soon.

So why not, for example, encourage schools to post lesson plans and homework assignments to a child's Facebook account? If children collaborate online about their homework assignment, so much the better. Most of what we do at work is collaborative. Why not encourage social education and make ebook widgets to enable this?

I recently got a chance to watch some college students studying for their finals. They came up with a new way of studying together by connecting over Skype and chat and sharing screenshots of the ebooks they were reading. What makes this interesting is that these weren't students studying together in the same dorm room or library but around the globe—in Dubai, Singapore, London, and Seattle. They cobbled together this setup themselves, without any help or guidance from professors.

It's important to worry about the future of education in a digital ebook-enabled world, especially if you have children, but I don't think the future's bleak. Instead, I think it's full of possibility. When I put on my futurist's hat, I see social connections everywhere inside ebooks. But even with all these social features, I think you'll be able to curl up with a familiar book and turn off all the naysayers and chitchatters in the margins of your book. You will always be able to turn off the popular highlights. You will always be able to unplug from the network and enjoy a book like you always did before, in a golden hour of sunshine with a great read.

Bookmark: Book Covers

There's a mysterious man on the subway. He's reading a book that you've read. There's something roguish or attractive about him that you can see in his face and in the way he carries himself, even though he's half hidden by the book he's reading. You're interested or maybe just tipsy enough to go over and talk to him. You casually point to that book he's reading, the one you've read too. And you start a conversation.

It's a scenario most of us have played out before, whether as the one approached or the one using the book as a pretext to get to know someone. Some of us have even met future husbands or wives this way.

The Spanish have a term for a chaperone who sometimes accompanies a couple on their first date: the chaperone is a *dueño* if it's a man, a *dueña* if it's a woman. The ebook revolution has killed the *dueño* of reading: the book cover. You'll no longer be able to have Gabriel García Márquez or Jane Austen chaperone you through your first hesitant and shy conversation as you talk as strangers about books you are reading, hoping perhaps for more intimacy or a longer conversation to get to know one another better. That's because book covers are already a casualty of the digital age.

Ebooks make a token concession to book covers in two ways. The first is by letting you see the cover on the web page where the ebooks are sold. The second is by often including the cover within the ebook itself. (However, some e-readers like Kindle skip right past the cover and go straight to where the book starts at chapter one.)

The demise of the book cover is a sad one, especially when you consider that many covers are works of art, as well as historical artifacts. Just consider the wild colors and bold lines of 1920s Russian book covers by Alexander Rodchenko, the lurid romance covers of the 1980s that featured Fabio, or even the way any book cover would fade to muted shades of blue if the book was out in the sun for too long in a storefront window. That's all gone now.

But you also have to consider that artistic book covers as we now know them are recent innovations. They've only been around for a hundred years. Before then, if a book had a cover at all, it was simply functional and undecorated, made to protect the book from excess wear and tear. At best, the covers would be gilt and hand-tooled from leather. They were symbolic encrustations of wealth rather than functions of advertising.

With digital books, though, you won't be able to catch a glimpse of the book that the airplane passenger sitting next to you is reading, so you won't be able to strike up a conversation quite as quickly. There's hope, though. I saw a recent tech innovation that lets you slip an iPhone into a special eInk sleeve so you can see images on both sides, and I thought it would be an amazing opportunity to show off book covers again, to beam the cover of the book you're reading onto the face of the device for everyone to see. And perhaps it won't be long before future tablets have glass screens on both sides that let you do the same thing. Maybe e-readers will start to show the book covers as screensavers. But there's a silver lining to the loss of book covers: the actual text of the book itself will come more to the forefront.

I can see a time when people will browse for books based on the content of the book, not the cover. Retailers will rank books for you based on the interior text. They'll automatically assess what a book's about and present the information to you when you need to make a purchasing decision. The loss of covers means that when you think back to an ebook you enjoyed, you'll perhaps recall more of the content of the book than the cover. You'll solidify more of the book's meaning in your mind rather than conjure up an image of the cover (which, by the way, is often created by a graphic designer who has never even read the book).

Still, for me at least, it's devastating how ebook covers are an appendix-like afterthought, tacked into books but rarely seen. At best, you see book covers on your e-reader's virtual bookshelf, but they're micro-sized and just a couple of pixels wide. I hate to say it, but I don't want to see book covers disappear! I'm almost tempted to wallpaper the inside of my home with book covers so I can be reminded of all my former books, all of them as familiar to me as friends. Because somehow, whenever I see a book with my mind's eye, I don't recall the text inside or abstract ideas it may have contained, but I do see the cover. For me, in a very real way, the cover *is* the book.

Am I alone in my appreciation of book covers? Let me know what you think about them, for good or for bad. And let me know what your favorite book cover is or any ideas you have for salvaging book covers in the digital age!

http://jasonmerkoski.com/eb/17.html

Libraries

Take a walk, if you will, through a university library, through one of the areas where nobody ever goes, like the section on 1870s foreign literature. Northwestern University outside of Chicago has a great library, and if you peruse its desolate dusty sections, you'll chance upon tomes from the era when books were bound with intricate marbled covers, a book-binding tradition that sadly is in decline. If you're lucky enough to find such a marbled book, you'll perhaps marvel at all its whorls and frothy bubbles, at all the inky emulsions! And the smells, the deliciously antique smell of old books, so musty, so brittle, so familiar but so sad.

A Kindle or iPad will never smell quite so lovely in its decline. If anything, it will smell of polyethylene and be frazzled like an overheated hair dryer. If it's white, it will take on the vaguely urine-colored tint that all old plastic gets when it ages.

But no e-reader will last as long as any book you'll find in a library. Kobos and Nooks and other devices will be relegated to sock drawers and trash bins, or lost in the garage sales and swap meets of techno-commodity fetishism. Devices like the Kindle have a lot of sales appeal, but only for a limited lifespan.

While the Kindle1 had such great demand in 2007 that it sold out in five hours and sold on eBay for 400 percent of the original price, it's doubtful that you could sell a Kindle1 today. There's always a later and greater device on the market. Companies who manufacture consumer products know this and design with this technical obsolescence in mind. As they're manufacturing the device that will hit the shelves tomorrow, they're already at work on its replacement.

While the reading hardware may age, the ebook content—being digital—is eternal. And likewise, because it's digital, it's possible to have a near-infinite number of copies of a given digital book. Perversely, though, your local library is only likely to have a handful of copies of a given digital book. Why is this?

Libraries have a fixed budget every year for what books they can purchase. So whether a given library wants to buy a print or a digital copy of a book, it's still going to have to pay for that book. That means that if you're late returning an ebook, you may still have to pay late fees (or, more humanely, the ebook will simply turn itself off and return to the library for another patron to use, even if you weren't finished reading it). This is because only a fixed number of patrons at a time can borrow the ebook from a library.

So even though digital inventory is infinite—even though all the patrons of the library could, in theory, download the same copy of the ebook at the same time, licensing terms will prohibit that from happening. Yes, you'll still have to reserve a digital ebook, just as you do a print book. The real benefit is that you'll be able to check out and download your library ebooks from anywhere. You're not going to need to go to a library to do that.

Libraries are always budget-constrained, and you're going to start seeing less and less shelf space dedicated to print books, because they're costly to maintain, rebind, stack, and insure. In an effort to save space and preserve shrinking resources, libraries will trend toward becoming miniature clouds of their own, collections of hard drives with all these ebooks on them. Perhaps the librarians themselves will become digital avatars of their former selves too, giving you online advice on which ebooks to read or which electronic encyclopedias and resources to use.

What will it mean as we lose this personal touch? Will we want to seek advice from an algorithm? Will we appreciate it if librarians are outsourced to a call center in the Philippines and there's no personal touch? No. I think we'll come to regret this loss. Whenever I visit my favorite library at the University of New Mexico, I find the librarians eager to help and anxious to please. I like the personal touch, the care they provide, and I believe we should work to embrace and preserve their crucial role as gatekeepers and conveyors of information.

I'm somewhat skeptical of the idea of digital librarians or digital libraries. Perhaps it's because some of the best years of my life were spent in libraries, surrounded by books. I really do love print books. I spent a lot of my childhood at the county library every Saturday, and I learned more from the MIT library than I ever did from my professors. I was so open in my reading attitudes that I would devour everything. Fiction, math books, history—it was all tasty. To this day, I still spend hours walking through the racks of libraries, poking through their basement stacks, and looking for interesting or esoteric tomes. Libraries offer a sense of discovery like no other.

Still, I was thrilled when my local library finally figured out how to offer ebooks to patrons. That night, I maxed out my library card and downloaded twenty books. The selection might still be small—only a few tens of thousands of ebooks—but I found abundant reading material. I ordered a pizza, stayed in, and read all night long—sheer bliss! And I love the convenience of being able to beam the books directly to my Kindle, instead of lugging them back from my local branch in the back of my pickup truck.

For me, it really is about books. They're not commodities, but soulful voices that actually speak to you. Some books whisper, some shout, and some seem to speak for no reason whatsoever. But I'm sensitive to the way they all sound, all these voices that stay mute until you open the covers and start reading.

I'm glad to see libraries embracing the promise of digital books, even though such books mean a threat of sorts to their continued existence—at least, the existence that libraries currently imagine for themselves. Because the charter of libraries is changing. Digital content is causing libraries to change now, just as newspapers changed ten years ago. For newspapers to thrive now, they have to target their local audiences. The ads need to be local, and so do the stories. Local papers can't maintain staff reporters to investigate events abroad anymore, and they don't need to. They focus on what's local.

Libraries can do the same. They can succeed by digitizing and making available local periodicals, historical archives, and books by regional authors. That's how they can differentiate themselves and stay afloat. In contrast, there's usually nothing local about a best-selling paperback. These more popular trade books are great candidates for being offered through a centralized, nationwide library service that local libraries can pay into.

As it stands now, individual libraries can sign up with a company called Overdrive to offer lendable ebooks—but many choose not to, for budget reasons. Having to provide print and digital books to patrons is a financial burden. I think the sooner we can accelerate the adoption of digital books, the better it will be for libraries and the more likely that some of the smaller libraries—often with great regional and local treasures—will survive into the decades ahead.

That said, I think there's one little-considered adjunct to libraries that will likely fade with the widespread adoption of ebooks, and that's the humble bookmobile.

On Main Street America, the bookmobile is as much a fixture as the ice cream truck, trundling down shady streets on summer afternoons, bringing library books to kids all over the country. In a digital age, it's hard to imagine a future for the bookmobile, except perhaps as an avant-garde piece of installation art from the past. It's not likely that the truck will drive down the streets letting the kids borrow digital books and download them onto their iPad minis, effectively zapping the children with ebooks.

In spite of the bookmobile's demise, libraries as a whole have a great future. I elaborate in the next "bookmark" about bookworms and how libraries are likely to become instrumental as cultural safeguards of books, as a check against rampant retailer sales practices and possible censorship. There's no better time than now to dust off your library card and check out some great ebooks to read on your iPad or Nook. You do have a library card, don't you? I've been using mine so much recently that I've memorized the twenty-digit bar code.

And I've fallen so much in love with my local library that I might just hug the librarian the next time I stop by.

For now, books can be preserved forever in digital form, like pressed violets between pages of an ebook in the cloud. As long as our ebooks can keep pace with changing file formats and are duplicated enough to avoid loss through hard-drive crashes, their future is assured.

The ebook revolution allows us, once and for all, to know ourselves. As a culture, we no longer need to fear death. The Constitution and Declaration of Independence will live on in digital form, even if the aging originals in Washington, DC, turn too brittle to read. We no longer need to fear culture loss—assuming, of course, that there's no futuristic form of library burning through selective viruses that attack a library's data center and preferentially wipe out ebooks, like digital Huns or Vandals.

Bookmark: Bookworms

Ebooks don't get viruses—not yet, anyway. Your own computer might succumb to a virus that turns it into a spambot zombie sending Viagra emails all over the globe or that monitors your keystrokes and sends your credit card numbers overseas. But your ebooks are safe. Until a nanovirus is made that can burrow through plastic and glass and eat away at resistors and diodes, the bookworm is an insect of the past.

I, for one, am glad I'll never have to see bookworms again. In the summer of 2000, I packed up all my belongings and put them in a storage facility in Boston before taking an international job assignment for a couple of years. Little did I know that just a few months after I left Boston, the storage facility would be flooded and my belongings on the basement floor pretty much ruined.

When I came back three years later in a moving van and saw the intricate colonies of fungus and rot on the walls, I was in despair. I opened up brittle cardboard boxes to find books whose pages were punctuated by insect tunnels and running lines of blue mold like antifreeze fluid. I lost hundreds of books, more than you'd find in an average public school library. It was devastating.

Culturally, though, we still face deterioration and loss of our content, although it's at the hands of something bigger than bookworm beetles. I'll put it to you like this: In the old days of antiquity, the works of Cicero and Plato were copied by hand, and because the copying took so long, scribes had to be choosy about what they preserved. If they didn't like a given book or didn't have enough parchment, they wouldn't copy it.

Because of this, we've lost a tragic number of works from antiquity.

- Aeschylus only survives in 10 percent of his seventy known works. The rest are now lost. Another playwright,

Sophocles, only survives into the twenty-first century with 5 percent of his works.

- Only half of Euclid's math books survive. Perhaps one of the missing books was an early work of calculus? If it had been more widely circulated in its day, maybe we'd have had computers by the Middle Ages and Greek colonies on the moon by now.
- Julius Caesar not only had time to defeat the French and become Rome's first emperor, but he also wrote fifteen books, of which only a third now survive.
- The Old Testament used to be much larger, with 46 percent of it now missing. As many as twenty-one lost books are referenced in the Bible (such as the Acts of Solomon and the Book of the Wars of the Lord). There are probably more that we don't even know about because the Bible never mentions them.
- Shakespeare fared better, with 93 percent of his works surviving, but even living as he did in the age of the printing press, at least three of his plays are lost, perhaps for good.

This same kind of blight can affect us with ebooks. If you take the long view of history and agree that wars and economic collapses and the redrawing of nations' lines will continue to happen, and that technologies will continue to shift, then it's inevitable that some of our ebooks will also one day become lost. But now, the magnitude for loss is much greater.

If a company like Google or Apple goes under, they might take all their books with them. It takes a lot to power a cloud, to keep all these ebooks humming in their hives. So in a large-scale book blight, there's as much chance that my aunt's book about her favorite cat will be preserved for posterity as that a book by J. K. Rowling might be. In fact, I would argue that an author's best strategy is to avoid making her works exclusive to any one retailer, that it's best to put your eggs in multiple baskets.

We can't read the future, but the opportunity for a wholesale book blight through negligence or gradual decay or decline in entertainment habits is greater now than ever before. Even if it's not caused by bookworms.

There are steps we could take to safeguard all our books from blight, of course. Libraries already excel in this respect, and as long as libraries continue to hold on to their own content and do not rely on the vaults of retailers, they can continue to help. In fact, there's an initiative called the Digital Public Library of America (DPLA) that aims to do this.

Led by a Harvard librarian, the DPLA aims to compete in a way with the Google Book project. Millions of volumes will still be digitized, but the libraries will be in charge, and independent readers worldwide can freely access digital holdings on their smartphones or computers. The DPLA is still in its early years, but its efforts—as well as those of similar projects, such as the World Digital Library project funded by the U.S. Library of Congress and UNESCO—may be the safeguards we need.

Librarians are unlikely heroes. Who would have thought that librarians would come to our culture's rescue, averting disaster and a literary bookocalypse?

That said, sadly, there isn't a single library from classical antiquity that has survived. I mention the top three such libraries in a later chapter, but there were other, smaller ones. They were all destroyed, with the possible exception of the personal library of Julius Caesar's father-in-law—and that only "survived" because it was buried under a hundred feet of hardened lava from an erupting volcano. About 1,800 scrolls "survive" in carbonized form. (Think of Han Solo frozen into a black block in *Star Wars*, or think of leaving a book at the site of an atom bomb explosion.) These scrolls aren't being read anytime soon.

We face the same problem of long-term survival with digitization efforts. Even if a book is digitized, will its file format survive? Will hardware even exist that can read it one day, centuries from now? Will the old Kindle or Nook in your desk drawer somehow survive the eons intact and surface again as a kind of Rosetta Stone that can be used to finally read and decipher troves of ebooks? Am I being too pessimistic in my worries for the future, or do you think we're not collectively worried enough about book blight?

http://jasonmerkoski.com/eb/18.html

The Future of E-Reader Hardware: Pico Projectors?

No doubt by now you've heard of Amazon's Microbook. It launched a few months ago, and being an early adopter, I was one of the first to buy it, try it out, and write a review.

The company describes the device in their promotional material as follows:

"The Microbook: An e-reader combined with a pico projector and connected to your Kindle account. No power cables. No hassles. No buttons."

The Microbook is very cheap because it has no screen and no moving parts.

It ships from Amazon's Japan offices, along with a little robot toy, although I'm not sure why. I can't read the instructions, but that's okay. As with any consumer electronics project, I shouldn't have to. It should just work.

All the Microbook needs is a network connection. My home's Wi-Fi worked just fine.

Because it was registered to me when I bought it, the Microbook knows who I am and what I'm currently reading. To read, all I need is a blank surface, like a wall or a table. So when I first turn the Microbook on and aim it at the wall, it shows the same page from the same book I'm currently reading on my Kindle.

There are no buttons, but it responds to voice control. "Turn the page," I say, and the image projected onto the wall changes to the next page. I can also tell it, "Go to the store," if I want to shop for ebooks.

Privacy is a bit of a problem, but I can read my books on the subway.

You can buy Microbook accessories, like a tripod for hands-free reading or a book with blank pages. This way you can pretend you're reading a print book.

What I like about it is that I can project the Microbook onto the ceiling at night when I read. It doesn't get too hot in my hands. And when I turn the Microbook off at night, the Japanese robot lights up its scary eyes.

There is no Microbook, of course. Not yet, anyway. I'm not aware of Amazon or any other retailer with plans for building such a device, but this is one of the ways I myself see the nature of e-readers changing.

When we hold a book or comic or magazine or even an e-reader in our hands, it's usually a flat object that is taller than it is wide. Most of the surface area is taken up with reading, with the content. But I think this is unnecessary. It's a waste of electricity to power such a large screen, and the objects are bulky. Besides, who wants to accidentally drop and crack an expensive iPad? I see a future when books can be projected with pico projectors onto walls, tables, and other surfaces.

There are great benefits to be found here. By projecting an ebook onto a surface, you're not constrained to a fixed size for the reading experience. The screen area could be as big or small as you prefer. Currently, you have to pay a premium for a larger device, whether it's an iPad over an iPad Mini or a Kindle DX over a regular Kindle.

Another benefit would be that your e-reader is very small and also cheaper, since most of what's involved in e-readers, in terms of hardware, is related to the screen. In fact, the screen itself is often the most expensive component of a dedicated e-reader, sometimes accounting for as much as half of the price. Get rid of the screen, and you can make a very small, very cheap device. Something perhaps the size of your thumbnail or a USB flash drive. All it would need is a network connection and a small pico projector.

A *pico projector* is an emerging technology that can beam large images from a miniscule machine. The word "pico" is used here in the sense of a picogram, a measurement the size of one trillionth of a gram. You'd perhaps unfold tiny tripod legs from the projector and aim it at a surface in front of you. Then you'd speak aloud the name of the book you want to read. If you don't already own it, you'd be prompted to buy it, at which point it would download onto this tiny device and start projecting. You'd navigate by voice commands, and the device would be cloud-powered.

This sort of device could socialize reading by making a book available to you and a close circle of friends. I can see this being used in reading groups, university study groups, or of course, in the privacy of your own bedroom. The biggest benefit of this type of device is its cost. Shrink the surface area of a device down to nothing, and you've made a cheap, hands-free reading device.

Of course, you can take this line of thinking further and make Nooks and Kindles really, really cheap. Make them so cheap that you give them away.

I foresee a time when Barnes & Noble, for example, will do just that. Perhaps at first they'll give Nooks away to people who buy a hundred dollars' worth of books a year. The retailer benefits because it saves on shipping, and it introduces new segments of its reading marketplace into the Nook experience. As programs like this are more and more successful, and as the manufacturing costs of these slimmed-down, cheaper Nooks drop, Barnes & Noble can afford to give them away to even more people for free. So now people who only spend seventy dollars a year on books—or even fifty dollars a year—can get a free Nook.

Over time, more and more people have Nooks. More and more people are reading. And that's good. Of course, these slimmed-down, cheaper Nooks are likely bare bones—no web browsing, no music or games, no bells, no whistles. But they're sufficient for reading itself and serve as a gateway drug to larger, more functional Nooks that Barnes & Noble sells for a steeper price.

If other ebook retailers use this model too, then e-readers will become very prevalent. You'll finally start to see e-readers in everyone's hands on subways, at bus stops, or during lunch breaks at work. If disposable e-readers become possible, you could get a new one in the mail every year, with newer features and better screens.

It's in Amazon's and everyone else's best interests to reduce the price of e-readers. This lets them increase the number of customers. Every twenty-dollar drop in price means legions of new customers for whom ebooks now become affordable. And in the final analysis, the logical price is free. Herbert Hoover once promised a chicken in every pot during the Depression, and in our own turbulent financial times, if e-readers are free, you'll find a Nook in every house—and if not a Nook, then a Sony or a Kindle or even an Apple device.

Just as e-readers are changing, so is the nature of how we read. I joked about privacy concerns with the Microbook. But it's not just people on the subway who can look over your shoulder at what you read. Retailers like Amazon or Apple can see every page turn and know every word you've highlighted, every annotation you've made. As you're reading on your beach chair, a giant looms behind you with a clipboard, peering over your shoulder.

Every so often as we read, our cloud-connected e-readers report back home to notify the retailer about where we are in the book. This is often done so that if you have multiple devices, you can sync all devices to the same page. But this also lets retailers like Amazon and Apple know how far you've read through a given ebook. They're able to monitor the progress you make. The information on reading patterns for a given ebook can be collected across multiple people, and the retailers can learn which ebooks were more successful. Do people abandon a given book halfway through? Is one particular chapter often skipped?

This information isn't yet being used to target your personal reading habits, but the reading patterns across multiple people for a given book could be used by the retailers—and sold back to publishers—to improve the quality of a given ebook. Perhaps the chapter that was often skipped needed to be better edited or needed an illustration to help explain what was going on. Or if the book is often abandoned partway through, then perhaps the publisher takes this information into account when it's time to renegotiate the author contract.

We're not yet at the point where ads will be targeted to you based on the paragraph or sentence you're currently reading, although Google does target ads to you if you've mentioned specific books, and Facebook's platform allows advertisers to do the same. Still, I think many readers are comfortable with this intrusion into their privacy, especially if it means better ebook prices. So I can imagine free ebooks that are 100 percent ad subsidized. You get these ebooks for free, but the catch is that on the bottom of every page, you see a contextual ad, perhaps based on the content on that page or perhaps based on your own web-surfing habits on the internet.

It's easy for retailers to serve ads to you across multiple websites. Ads are sticky, like cockroaches dipped in honey, and you can't quite get rid of them. It may not be long before you start to see those same sticky ads following you around on your ebooks. But until then, I think that your reading privacy will only be bent to provide statistics back to publishers, in the way I just outlined. If this ultimately serves to make for better-designed ebooks, and we as readers are oblivious to the way this data is being used, then perhaps there's no harm to us in the process. For now.

I mention the pico-projector e-reader as an example of the kind of disruptive hardware technology we may see in the years ahead. The future of ebooks is just getting started, after all. Lots of new technology will be coming to e-readers. Some types use organic crystals woven into intricate patterns or arranged in spirals. Some work like the wings on a butterfly, reflecting light at the right frequencies to reproduce full color. E-reader technology is an area of ongoing innovation, and the devices that will be out in just a couple of years will make existing eInk displays look like Edison's wax cylinders.

Eventually, e-readers may get so cheap that they're unprofitable for retailers to sell. That makes you wonder whether they'll continue to be sold. But consider the history of razor blades.

In 1895, an inventor named King Gillette turned away from his architectural drawings of futuristic cities and utopias and hit on an idea for a new kind of razor. It took him ten years before he could manufacture them, but they were revolutionary. Instead of having to buy a razor blade and sharpen it before every use, you could buy reusable razor handles and disposable steel razor blades from Gillette. When the blades got dull, you just bought new ones. Gillette took a loss on every razor handle he sold. But what good is a handle without a blade? None at all, so he made a tidy profit on every blade he sold.

Using this as a metaphor, you might ask whether a given retailer is in business to sell razors or blades. I say the answer is both. In truth, ebooks and e-readers are part of the flywheel for any ebook retailer. You can't sell content without a reader, and you can't sell readers without content, so you need both.

No company can rest on its laurels yet and just focus on ebook content while letting the others provide the readers. Even Google ended up launching its own smartphone and tablet. Although there are challenges and pains that you go through as an organization to build out a device and the profit margin may not be high, by owning the reading experience at the hardware level, you can do things with content that no one else can.

Recent events have already shown that a retailer will take a loss on hardware if the content can make up for it in sales. For example, when Amazon released the Kindle Fire, an inexpensive tablet that could compete with Apple's iPad, many manufacturing pundits believed that Amazon was losing money on every Kindle Fire it sold as a piece of hardware but was making up the balance on content sales. It was a brilliant business decision later mirrored by inexpensive e-reader tablets from Amazon's competitors.

The same will hold true for future e-readers. If anything, prices will drop to levels so scary that corporate accountants and decision-makers at major retailers will need to have nerves of steel.

Bookmark: Lost Libraries

In doing research for this book, I wanted to watch old TV episodes of *Oprah* to find the day when Oprah discussed the Kindle with Jeff Bezos. It was a pivotal day for Kindle. Based on her show, the original Kindle sold out forever. In its way, the interview between Jeff and Oprah was a unique moment in history—for books, anyway. Between the two of them, Jeff and Oprah had done more than anyone else to promote and sell books in this century. You'd have to go back a hundred years to find another person who singlehandedly had as much impact on reading, and that was Andrew Carnegie, who opened 2,500 free libraries around the country at a time when American libraries were closed to the public.

But just a mere two years after the *Oprah* show aired, it's no longer available anywhere on the internet or even the undernet. The show had a daily viewership in the millions, but it isn't available anymore, with the exception of occasional bootleg clips here and there, like bits of papyrus buried in the Egyptian desert of the internet.

Media has a surprisingly short shelf life. For example, only four of the films of Theda Bara survive. The others are all gone, lost. Theda Bara was the original Hollywood vamp, one of the most massively popular actresses in all of movie history. In 1917, her film *Cleopatra* had the biggest budget of any film up to that point, $500,000, and that was at the end of World War I! And yet, all that remains of the film and Theda's risqué outfits is a smudgy, five-second clip that was rescued from the vaults of the film studio as it was burning down decades later.

Theda Bara isn't unique in this sense. Only ninety seconds of footage exists from one of the first animated movies, *The Centaurs*, made ten years before Disney came onto the scene. One of the first Westerns, *Devil Dog Dawson*, only survives as a thirty-eight-second fragment, found by accident in a mislabeled film can in Ohio. The first Technicolor film, *On with the Show*, a crowning success that raked in the modern equivalent of $2 billion in revenues, is now completely lost, although somehow, absurdly, a twenty-second color clip was found in a toy projector in the 1970s.

History was harsh with Theda Bara and a lot of other silent film stars, but it's just as harsh with books. If you look back to the ancient world, there were three major libraries. First and foremost was the library of Alexandria in Egypt with about half a million volumes, then the library of Pergamum in Greece with 200,000 books, and then finally the library of Harran in Turkey. These three libraries held most of the books of the ancient world, and scholars still gnash their teeth and tear out their hair thinking about all the conquerors in the intervening centuries who dumped these books into rivers or burned them for fuel.

The story of books in the ancient world is a sad one. Anthony dismantled the library of Pergamum as a wedding present to Cleopatra. He emptied the shelves and sent all the books to Alexandria. But that library didn't last long, because it was repeatedly decimated by fires and finally Islamic conquest. The only sizeable collection of books from the ancient world survived in Harran, a dusty outpost in Turkey where all the scholars fled from Egypt and Greece with their books. The books stayed hidden there until Arab scholars rediscovered and retranslated them, leading in part to the Renaissance of knowledge around Gutenberg's time.

The fate of these ancient libraries is instructive and offers models of what might happen with corporate mergers and ebooks. Is it too hard to imagine a future where Google and Apple merge and combine their vast ebook libraries—only to suffer the slings and arrows of corporate fortune and go bankrupt one day, the books disappearing as the servers get shut down and rust, as distant data centers become overgrown with ivy and vines? Perhaps Amazon survives for a while before it, in turn, is acquired by some future new-media company, its ebooks relegated to an archive, perhaps to survive, perhaps not.

What would it be like to live in a future where all media is consolidated under one company? Not only would that company be able to set arbitrarily high prices on content, but it could also bury any content in its vaults, effectively censoring it. And what would it be like if that company failed, went bankrupt, or worse, lost its media archives? What if all the content was destroyed, perhaps through a massive server outage or an act of internal sabotage by a disgruntled employee or a digital ebook-eating virus?

Such a loss is too catastrophic to consider. But it could happen. Technological obsolescence not only happens to hardware and software, but also to institutions. After all, there were only three major libraries in the ancient world—and only one of them survived long enough for its books to be retranslated and preserved. Likewise, there are only three major digital media retailers now—Apple, Amazon, and Google. Which of these three, if any, do you think will survive? Fast-forward a hundred years: what do you think it would be like if one company monopolized our media?

http://jasonmerkoski.com/eb/19.html

The Future of Writing

The print revolution in Gutenberg's time was truly revolutionary because it allowed knowledge to be distributed to masses of people. It was no longer necessary to hoard parchment, and books weren't only available for the elite. Printing has undergone changes since then, but most of them have been evolutionary rather than *rev*olutionary.

For example, when mass-market paperbacks emerged in the mid-1930s, they weren't revolutionary. Mass-market paperbacks were pioneered by the Penguin publishing house, which took the novel approach of producing books from cheap pulp, hence the term "pulp fiction." In fact, the mass-market paperback books themselves could be recycled into pulp and reused as paper for the publisher's next mass-market paperback. All the books you see in grocery-store checkout aisles and airports owe their existence to the mass-market paperback format. The idea was evolutionary because it allowed books to be sold for even cheaper prices and for incrementally more people to read them.

Don't get me wrong; we need evolutionary improvements.

But revolutions are acts of genius. They take multiple evolutionary improvements and compress them into one new product. Gutenberg's printing was revolutionary because it combined multiple evolutionary improvements (moveable type, the printing press, and oil-based inks). The iPhone was revolutionary in the same way (large touchscreen phone, apps, GPS, and unlimited data plans), and so were ebooks.

As a culture, we can't go back to the pre-iPhone days of the mere cell phone. And we can't go back to the pre-ebook days of Borders, B. Dalton, and your local bookseller. In part, that's because these stores are closed, bankrupted. The immediacy of digital ebook downloads and the convenience of a cloud-based library have replaced them. Moreover, ebooks are eternal.

Classical scholars may hope one day to find a lost work of Aeschylus in the bindings of an Egyptian mummy or Shakespeare's *Love's Labors Won* in an old English priory. But ebooks democratize and extend the longevity of books. Your aunt's self-published volume of cat poetry will survive the eons, and your grandpa's autobiography will help your descendant in the twenty-fourth century to build a family tree. Our words aren't dependent on penurious scribes or budget-minded librarians or choosy auditors at the Library of Congress. Our words are liberated—that is, if we choose to write them in the first place.

Paradoxically, even though ebooks have ushered in a revolution in reading, the digital culture of our internet age is making writing more difficult.

The flip side of digital reading is digital creation. I've written a lot here about how ebooks are changing the way we read, but how is digital technology changing how we write?

We're getting more used to the idea of ebooks, but many artists still prefer to sketch on paper instead of the digital medium, and many writers still prefer to carry journals with them in which to record their ideas and impressions.

Digital journals still haven't become mainstream. However, some companies are providing an interesting bridge between print and digital for writing. For example, Moleskine and Evernote have recently partnered to create a hybrid system that lets you write or draw on a special kind of paper inside Moleskine notebooks. The paper lends itself to automatic digitization and cloud upload through Evernote, a company that aims to let you archive and revisit all your memos and ideas online. I think this is great because it can make content more searchable and reusable. Content from a journal can be copied and pasted into a term paper or business plan rather than having to be retyped. Innovations like this make us more efficient.

I'd like to say that I'm an early adopter and I'm deliberately choosing to write 100 percent digitally from now on, on principle. But the reality is otherwise—I had to learn from my own misfortune to go fully digital.

Sometime over the weekend in summer, on a rare vacation, I lost my own journal. It might have been at a bakery or a farmers' market. It might have been at a bar or a gallery, but I lost it. I no longer have my journal.

It was a blue journal, the size of a regular notebook, with drawings and writing inside. It's worthless to anyone but me. It has everything I wrote over the last two years. There are no passwords or bank account numbers in my journal. But there are illustrations I made and ideas I had about digital media.

What is the lesson of losing my journal?

I need to go fully digital.

I learned that you can't back up your journal to Dropbox or any other cloud backup service. And who knew? There's no kiosk in a mall where you can back up or scan in regular, everyday objects so you can restore them if you lose them. There would be such a service in the ideal world, in the best of all possible worlds, but there isn't here. Dropbox is great, but it won't work for real-world artifacts, for things. Only bits.

Because of the experience of losing my journal, I decided to go fully digital in my writing from now on. The experience of losing my journal has turned me into that guy in the corner of Starbucks who wears headphones and talks to his iPad, dictating his thoughts in a public place and making a fool of himself. It has exiled me to the shady corners of coffee shops and bus stations, away from the happy patrons who I might otherwise annoy with my dictational monologue. Losing my journal has made me mutter to myself like an old, drunken crazy man.

There's a benefit to this, of course. I can now back up whatever I write. If I have a kid one day, I'll give her all my blank journals and let her draw in them. Or maybe I'll give my kid a secondhand iPad, so her scrawls last a thousand digital years. The digital mode of creation immortalizes us; the analog mode humbles us.

I'm not the first writer in history to lose a journal. Some writers have lost more. Malcolm Lowry, author of *Under the Volcano*, retired to the coast of British Columbia to write his second novel. He spent seven years writing in a cabin he built out of driftwood, but when he was about to mail his manuscript to his publisher, his house burned down and he lost everything. He had to spend another seven years rewriting and reconstructing the book, which was finally published as *Ultramarine* but which even the author himself admitted was a bit of a flop. Something about the loss affected his writing, and the book was never as good as the original. Reconstructions rarely are.

There's something about the surety of the Save button. When I save a document—like this one, as I type it on my computer—I know it's instantly copied to the cloud. I'm safe, my writings are backed up, and I rest easy. Until, at least, the cloud topples over one day, and what was once in the much-vaunted cloud is reduced to digital dust. I imagine a colossal implosion, like that in the season finale of *Lost* where Locke destroys the Dharma Initiative's Swan station. In the buildup to the countdown on *Lost*, you see computers toppling over, steel walls imploding, all the girders creaking and straining, the countdown clock itself imploding, knives flying, metal struts rending and shrieking, and you hear the high-pitched whine of metal, the catastrophonic sounds of electromagnets bending the walls, and a woman's voice announcing systems failure, before Locke admits, "I was wrong." Easily the best three minutes in TV history.

But the digital cloud won't topple for a while.

Maybe it will in fifty years or so, once we can no longer afford to power the many clouds. Facebook has its cloud, and so do Amazon, Apple, and Twitter. Basically, it's new gold-rush country, but instead of gold, people are mining clouds. Old stalwart companies like Adobe and even Walmart need to have their own clouds. There are even companies selling devices to small businesses so they can manage their own clouds.

It's faddish, and it may all fail one day.

Maybe then people will return to a simpler mode of life. Return to writing with pen and ink and, yes, the peril of permanently losing what you've written.

Both print and digital are ephemeral. Our works can be destroyed in an instant with either. But at least digital versions give you backups.

All these backups do introduce one casualty, however. With digital writing, there will be fewer manuscripts for sale. There used to be a healthy after-market among collectors to buy not only the first editions of a given novel, but also the author's own manuscript. With digital manuscripts, this kind of collecting is pointless. Since value is typically related to scarcity, there's no value to a unique good if it can be duplicated an infinite number of times.

Though I'm bullish on digital writing, I do want to note that in the process of writing this chapter, my word-processing software crashed. Twice. I almost lost everything I wrote. Even after recovering what I could, a lot had to be rewritten from memory. So take this paean to the benefits of digital writing with a grain of digital salt!

That caveat aside, now that so much is digitally composed, authorship is flourishing.

Writing digitally instead of on old-fashioned typewriters lends itself to faster publication. Ebooks can be self-published in just hours. Retailers like Barnes & Noble and Amazon and Smashwords, who have their own self-publishing portals, have created ways of disconnecting authors from publishers. Authors who are savvy enough to use the newer self-publishing tools are flourishing.

It's been said of the American Revolution, in Lincoln's Gettysburg Address, that it was a revolution of the people, for the people, and by the people. But let me tell you something about the ebook revolution: this is a revolution of the publishers, for the readers, by the retailers.

Strangely enough, the roles of authors are mostly unchanged in this revolution. They're still going to write on their computers or typewriters. Yes, some manuscripts are still typed out on old typewriters, although composing books digitally lends them a felicitous fluidity that enables them to be published faster. They're more immediate now than ever before.

In fact, digital authorship is alive and well! So many wonderfully obscure and unknown books are published on Kindle's self-publishing platform, ranging from "lost books" of the *Odyssey* to bizarre theories of the universe. The rise of self-publishing and ebooks is giving little-known authors a megaphone. And sometimes the authors are heard. A few of them start out as self-published and then resell their books to big publishers, where they actually can be more successful.

Such authors may have gotten their start with self-publishing, but they become truly successful only when courted afterward by a traditional publisher, who helps the author craft the raw content into a polished product that readers really want. Traditional publishers also provide such initially self-published authors with a bigger and better marketing platform, a broader reach, and frankly, legitimacy.

I joke sometimes that self-publishing is mainly about cat poetry, but the universe of self-published content is truly vast. If only it were possible to surf through all these words easily, if only I could have a buffet of this content all at once—and truly, I do love to read, because to me reading is like being a starving omnivore at an all-you-can-eat buffet!

There's no better time than now to be a reader.

And it's a truly democratic time now for being an author, in that anyone who can use Microsoft Word or blog-authoring tools can quickly publish a book—but with democracy comes altogether too many options. In some ways, this is a case where the paradox of choice reigns supreme. Two or three hundred years ago, your choice of authors might've been limited to Daniel Defoe or Jonathan Swift, but now there are too many authors, too many *choices*. Almost too many choices, in fact.

There are so many choices now that we may be afraid to make a choice. This is known as the *paradox of choice*. It's easier to choose between chocolate and vanilla than to choose between fifty-seven flavors at your local Baskin-Robbins. You'll stare at all the flavors, numbed and likely to leave in despair, overwhelmed by all the choices.

In addition to swamping readers with a surfeit of great books, the ebook revolution will place demands on authors. Especially authors who choose to use larger publishers.

Publishers will eventually require their authors to log in to websites that show statistics about their books. For a given chapter that the author wrote, the sites will show statistics about what percent of people read that chapter, which pages were highlighted the most, which pages were shared most on social networks, and spelling mistakes or anachronisms that readers pointed out.

The author will need to use this data in making revisions for a second or third edition of the book. Or perhaps this data will be available to the author as he or she plans a new book—to see what content engaged readers most and what sections were too difficult for the target audience to read. Editors may not even be part of the process. Authorship may be a direct relationship between readers and authors, mediated by these web pages of statistics culled from the thousands of readers and their reactions to the content.

In the years ahead, authors will have to become amateur statisticians.

Likewise, these stats will open up worlds of possibility for the writing process itself. It will mean that the process of writing an ebook is no longer static.

When traditional brands launch a new ad campaign, they usually create multiple versions of the ad for what's called "A/B" testing. Version A of the ad gets shown to one group of people online in a test market, and version B gets shown to another group. After a few days, the results are tallied, and whichever ad is more effective is picked to go national. The same process will happen with ebooks. An author will be able to publish version A of an ebook with a plotline that differs slightly from version B. Once results are in from readers, the author can pick the more effective ebook.

The ebook you read may be different from the ebook your friend reads, even if it has the same title. Ebooks are no longer static, in the same way that your experience of a given website is different from my own because the ads are customized, different for each of us. For an author, writing an ebook will be like a visit to the optometrist and getting fitted for new glasses. Better A? Better B?

The future of writing is changing and requires authors to be part engineer, part marketer, part statistician, and oh yes, part writer too.

Bookmark: Degraded Text

Monks in medieval monasteries worked hard to preserve the writing in scrolls and parchments whenever they'd recopy them. The same thing is true of the Arabs who cared for most of the Greek texts that we now have from the era of Plato and Sophocles. They were very wary of making textual missteps and introducing errors, concerned as they were with the wisdom of the past and the purity of the written word.

But now we're in a more capitalistic time. There's a frenzy of digitization going on with ebooks, and everyone's rushing out of the woodwork to profit from books. As these profiteers rush to the market, the texts that they're bringing us are getting degraded.

For example, there's a wholesale rush to take all the books in free web archives that have been around for decades—like Project Gutenberg—and repackage them for Kindle and iPad. People are doing these conversions because such books, while not as popular as bestsellers from *The New York Times*, still have fans who would pay a modest premium to buy such a book. Although these books are likely to garner fewer sales than a more contemporary, popular book, a lot less effort is needed to ready them for sale to consumers. After all, the book has already been written and already digitized. It just needs to be reconverted to ebook format.

Because there's such a frenzy happening now, however, many people are converting the same books over and over again and offering multiple versions of the same book for sale. Sometimes scores of different versions of the same text exist. But each version is corrupted in different ways. Because people are processing these books in bulk, they are rarely edited, and conversion errors rarely get caught. As a result, the books often get adulterated, and sections are chopped up or lost.

In extreme cases, the book is no longer readable. More frequently, explosions of bizarre symbols from the outer reaches of your keyboard appear in the text for no reason whatsoever, like someone swearing at you in a foreign language. There's a great Japanese word for this: *mojibake*. This phenomenon is often seen on international web pages when they're rendered badly in browsers, but it's happening a lot in ebooks too. While this problem is usually more noticeable in public domain content, you'll sometimes see it even in books from top-tier publishers, such as when an ebook is rushed to publication with insufficient time to review the quality.

You can't blame the top publishers for moving fast. In a way, everyone involved with publishing is a bit of a hustler these days. There's a wealth of opportunity to digitize the world's content, an opportunity that will only come once to our culture as a whole.

Some of these hustlers and opportunists are very crafty. I know a Russian guy who took a flatbed scanner into the Kremlin archives and scanned away relentlessly at all the books in the archives, with the intent of selling them digitally. After doing this for three years, he had enough books to sell as cheap print reproductions of the originals, but he couldn't convert them into digital books. That's because his scanners only did black and white, with none of the shades of gray in between, and the quality was just too poor to convert the scans into digital books.

There's a whole spectrum of opportunity for would-be *mojibake* hustlers these days. So go out and get yourself a flatbed scanner, fly to Iceland or Norway for a couple of years, and see what you can digitize! The gold rush is on to digitize content, and while you're prospecting for gold, you might find entirely new minerals that no one is even aware of and for which there soon will be a market.

I'm thinking in particular about sheet music scores, old pamphlets, or postcards. There's a wealth of material to scan in and digitize, and books are just one part of this. Newspapers and magazines are part of this gold rush, although frankly—and I'm completely unbiased here—books are sexier than anything else.

In addition to books, countless pamphlets, comics, newspapers, 'zines, and ephemera of every age could be digitized and made available. Books are only the surface of what's possible. The printed word goes much deeper than the surface, and there's a vast shadowy biosphere of words that's currently unexplored and undigitized.

But what do you think? What would you really like to see digitized? What ephemera from our print past—from cereal boxes to greeting cards—do you think needs to be preserved for perpetuity?

http://jasonmerkoski.com/eb/20.html

Digitizing Culture

There are a lot of different versions of how the future could play out in ebooks, but what I see happening first is what I'll call the "utility" model, which is kind of like having ebooks available under a monthly Netflix-like subscription. We view electricity and water and TV as utilities, and most of us have to subscribe to them. Some are flat-fee, and some are priced based on how much they're used.

Right now, when you buy an ebook, you're making a one-time transaction. But in the utility model, you would pay one monthly or yearly lump sum to get unlimited downloads. Perhaps the books wouldn't actually be yours; think of them as rentals, available whenever you want to read. The download happens as fast as always, and the ebook is on your e-reader for you to read. Perhaps it expires in a week or two, but you can always download it again. It's like a faucet. Water comes out of the faucet when you turn it on, and thus ebooks will, as well.

Amazon recently launched such a Netflix for ebooks, but only a very few books from the Amazon catalog are part of it, and you only get one free ebook a month. Would you use Netflix if it only showed nature documentaries, episodes of the 1980s cartoon *He-Man*, and Mexican wrestling matches? Unless you're an aficionado of *lucha libre*, you'll probably wait until more ebooks are part of the program.

Of course, true book lovers may cling to their print books. For one, books smell nice—although it's possible that e-reader manufacturers could add the “old book” smell to their products. It’s been shown that volatile chemicals like acetic acid, furfural, and lipid peroxides contribute to that musty smell, and they could easily be swirled into the e-reader’s plastic when it’s being manufactured.

Another reason why book lovers aren’t giving up print books is because the books are not yet available electronically. When Kindle launched, about 90,000 ebooks were available for sale on its website. Even now, at the time of this book’s writing, there are only 1.8 million Kindle ebooks. This may sound like a lot, but it’s chump change compared to the 35 million books in print. Early adopters like CEOs and former presidents and astronauts would have no problem buying e-readers despite the limited content selection, but mainstream readers demand more content selection.

To get even more books digitized, I think a company could make a device the size of a toaster oven into which you could put a book, a device that would work on books of most sizes. The device would page quickly through the book, take a picture of each page, and upload the pages to the cloud. The toasters would have to be intelligent enough to correct for poor lighting conditions and the way the words get elongated near the break between pages. If you don’t believe me, try placing a book on a photocopier and see what you get. See how the words get distorted and unreadable near the spine.

I imagine it would be called the “ebook toaster.” As far as gadgets go, the ebook toaster would be kind of dangerous. A warning label on it would recommend that it be used by people age 18 and older. Why? The two blades inside the toaster would be sharp enough to slice off the spine of a book. And just as a regular toaster has a tray to catch the breadcrumbs from your bagels or pizzas, the ebook toaster would have a tray to catch the spine that’s been snipped. If books could bleed, this tray would catch the blood.

Mechanical robot arms would unfurl themselves within the toaster and lift each page, one at a time. An arrangement of mirrors and cameras would carefully take pictures of each page. When that process is done, you would either put a rubber band around the remaining pages to keep the print book, or you’d dispose of it. As long as your ebook toaster is connected to your home’s Wi-Fi network, you would get your ebook back in about an hour. It would show up on your e-reader's home screen, ready to read, with no crumbs or burnt crusts.

When the ebook toaster finished its job, the ebook would be reassembled in a reflowable format from each of the original book's pages. This would allow you to convert your library of print books to digital. Perhaps it would take a half hour per book, but once it was done, the process would be like ripping a CD into MP3 files. You'd have the ebook files accessible anytime, anywhere, regardless of device. In 2003, I spent a few months slowly inserting all my CDs into my computer to gradually digitize my music collection, and now I have those music files forever.

Perhaps instead of doing this at home with an ebook toaster, you'd hire a company to do the conversion for you. Readers like you and me aren't going to sign contracts with conversion houses in India or the Philippines to convert our personal libraries, book by book. But that's okay. Companies will come into existence to do this for you, at a cost. I think that in a few years, you'll be able to mail boxes of your print books to conversion facilities that will manage the print-to-digital conversion and send you back files in the format of your choice. Or maybe you'll see larger versions of the ebook toasters at the mall.

You may see kiosks in the mall where you can bring your books and get them converted. You can watch while they do it, or come back after you've gotten a pretzel from the food court, and then collect your digital books on your flash drive. You'll see these kiosks in malls, and probably small stores too, where retail space is cheap enough. They'll be a lot like eBay shipping centers where you go to get your goods packaged and mailed, a service that you're happy to pay a small surcharge for, to save yourself the hassle.

It will be a lot like going to the mechanic to get the tires on your car changed, except that now you will have newer, better tires. And yes, you'll still have to pay a handling fee to dispose of the old tires or, in this case, your print books. The conversion machines will likely use what's called destructive scanning, meaning that the book has to be destroyed to be converted. This is what most major publishers do when they have a print book that they want to convert into digital format.

When I've traveled to destructive scanning facilities, I've seen machines that seem like they belong in a slaughterhouse, machines with whirling knives that slice the spine straight off the back of the book. Sometimes the process is more manual and less sophisticated. It may be a team of women in India sitting at a long table, holding razor blades, and doing the same work, but much more cheaply.

I think you'll see such a process at the mall, where nimble-fingered teens wield razor blades to scrape the spines from your book so that they can quickly scan each individual page. The book will be destroyed in the process, but the process will be painless for you—unless you had any emotional attachment to the book. It will be like a visit to LensCrafters, where you get your new glasses in about an hour.

You can easily imagine the shady file-sharing markets that might emerge as people learn that they can swap these scanned-in files with one another. Or maybe people will go to bookstores with these toaster-sized devices under their trench coats and scan in this week's bestsellers. But in a positive sense, I think this type of conversion will help the used ebook market grow, making that eventuality turn into an inevitability. Maybe with this kind of device, legitimate used ebook stores will emerge. Maybe used ebooks can be resold once or twice before they spontaneously combust like Maxwell Smart's secret messages.

Books are important, so let the consumers have them, used or otherwise. Publishers should get a fair price, as should authors and any middlemen like retailers, without whom the entire ecosystem would fail. Likewise, I think libraries can benefit. There might even be a company whose sole purpose would be to allow libraries to exchange digital copies of one another's scanned books so that they don't have to rescan each book at each library.

The value of books will change, of course, and perhaps for the better. Right now, books that are esoteric and hard to find are at a premium because there are few print copies of them. But once a book is digitized, with endless amounts of secure backups, there's no reason why prices shouldn't drop. And prices should follow a new paradigm: the price of a book should be inversely proportional to its popularity.

We see this now with out-of-print books from before 1923. When digitized, they're commonly free. They're part of the public domain. There are older books that are not part of the public domain, not yet, and when they're digitized, they'll be of interest to historians and scholars and anyone who happens to follow links to them in a possible Facebook for Books. The cost of these older books should be damn cheap, almost zero.

Conversely, the most popular books of the day—like those on *The New York Times* bestseller list—should be at a premium, in keeping with the marketing investment that the publishers spent to promote them and create consumer demand. But a book that was on *The New York Times* list five years ago is rarely worth the same as what's on the list this week. We see the decay in price of new titles, but older, rarer books are still inflated in price because they haven't been digitized.

There's a chilling reversal, though, by which retailers might become the new libraries.

This is a scary mind shift, but it is in keeping with the currents of our culture as we commoditize every aspect of our lives. Given these currents, it makes sense that retailers will assume stewardship of our culture. Libraries once held all of the world's knowledge, but, with rare exceptions, there is no longer any library on the planet with a larger collection than the books currently held by the likes of Amazon or Google or Barnes & Noble. Information is available, but it's no longer freely available.

This is a future that I don't entirely welcome for philosophical reasons, but it does seem likely. Retailers might become the new libraries. Perhaps this happens first by publishers acquiring one another so that they can lobby for favorable ebook terms and discounts with retailers. Indeed, we're already seeing this, with the recent merger between Random House and Penguin. To be competitive, smaller publishers may feel pressure to acquire other publishers or merge with them so that, as a bloc, they can negotiate with the retailers.

Eventually, though, what's to stop a company like Amazon from acquiring one of these large publisher conglomerates? Apple might then have to retaliate and buy another mega-publisher. Retailers will try to acquire publishers' vast content holdings in a bid to become the predominant purveyor of the written word—whether in book form, magazine form, or pamphlet form.

And once this future is played out, then what happens? Do the retailers themselves converge and consolidate, like banks did in the 1990s? Are they acquired by the governments, in response to the monopolization of the written word or because of fears that retailers will hijack the language itself and censor it? Does Apple send emissaries out to all the state libraries of the world and license digital rights to their content?

I can't tell you. My crystal ball is dark regarding this matter. When I first joined Amazon, they gave me a Magic 8 Ball. They gave them to all new employees at the time. When I shake my Amazon-issued Magic 8 Ball, this time it says, "Ask Again Later." For now anyway, the future is as cloudy and as dark as a busted eInk screen.

Only one thing is certain: content was, and is, still king.

Clearly, as a culture, we're smitten with the digital. Ebooks make sense for so many reasons. But what will happen to print books in the years ahead?

As more and more people buy ebooks, they'll start to preferentially buy ebooks, because the experience is so "sticky" and because the more digital books you have, the more you gain from the network effects of searching and indexing, something that works poorly in print books.

Eventually, there will be a tipping point at which the benefits of digital outweigh print, and there will be a mass shift from print to digital. Nobody in the book industry is sure exactly where the tipping point is for selection. It may be that Amazon or Apple needs to have 95 percent digital coverage of all the books in print before people stampede to ebooks.

But for a time, people will have libraries that are part digital, part print. Those of us who see what's coming realize that as more consumers start buying ebooks, they're going to look at their personal libraries of print books and try to figure out what to do with them, since they're becoming obsolete.

The obvious thing to do is to sell them.

You're going to see a lot more used book sales in the next ten years than ever before. People are going to start dumping their print books to get whatever prices they can from them, simply because it's more convenient to go digital. And let me tell you, print books are not convenient. I have four thousand of them, which means that every time I move into a new house, I have to box them up and haul them around, something my back may not be able to handle one day! So I've started selling them.

I've sold more than a thousand of my print books already on Amazon's used book marketplace. It's simple to do, especially if you have a computer with a video camera on it. The video camera can scan in the bar codes on the back of the books through the same process that retail stores use to scan products with a laser at checkout. Software like Delicious Library automates a lot of this, and you can often get a free membership to Amazon's Seller Central that lets you sell your used books. You don't have to work at Amazon like I did to get these benefits!

There's actually a thriving subculture of people armed with laptops who go to used bookstores, scan in the bar codes with their video cameras, and see if any of the books are worth enough to buy used from the bookstore. If so, they resell the book online at a higher price. You'll see more of this in the years ahead, as well as better tools on smartphones to allow non-experts to make a living at this.

I think we're going to see a huge number of used book sales in the next five years as digital books go more mainstream. We may have a glut of books, in fact. If you're like me and you think that the moldering, forlorn used books for sale on a scrappy rack outside a bookstore are sad, then wait until you see how much sadder it will get for printed books.

Books that can fetch ten to twenty dollars today in used condition will be lucky to sell for ten to twenty cents in a few years, simply because the market for print books will be flooded and no one will need them anymore. A few book collectors will snap up the choicest pickings offered for sale, but the great majority won't get sold, even at a penny each, because the buyers simply won't exist in great enough numbers. It will be a buyer's market. Unsold books will get donated to libraries, but even libraries don't have infinite amounts of space.

With this glut of used books will come a perceived cheapening of books. Our culture still perceives print books as precious, even in the age of mass production. Books are still seen as status symbols, after all.

The wealthy often have finely apportioned libraries in their mansions (even if they're often just decorative).

But what will our culture be like when people start dumping their books because they simply can't sell them? You'll first see piles of unsold books outside the hipster neighborhoods in New York and San Francisco; then you'll see piles of books by the trash on Sunday night for anyone to take. Then you'll see community events in other cities where people get together on weekends and swap books.

If there are 119 million readers in the United States, and every reader has an average of one hundred books, and half of these will be eliminated over the next ten years as people go digital, about six billion books will need to be disposed of in some way or another. That's around four billion tons, or the equivalent of ten years of trash. It all has to end up somewhere.

I think you'll see a strange hybrid between dump and library—perhaps a section of the dump for books that's cordoned off from the trash, a section run by profiteers or book-loving volunteers who will sell books by the pound or perhaps the truckload. What will the books be used for? Perhaps firewood or fuel.

As books move to digital and fewer are printed, fewer print books will be sold, which will hasten the decline of bookstores. Physical brick-and-mortar bookstores are already struggling to compete with online giants like Amazon or Walmart when it comes to price and selection, and the move to digital just hastens the already sad decline.

Most retail bookstores haven't made the transition to the new digital culture. Even in university towns, the kinds of places where people usually read more, bookstores are closing their doors. Conventional bookstores that have focused on offering new books will fold, hit with the double whammy of fewer books being printed and the glut of used books on the market.

Publishers often complain that digital books are forcing their hardcovers to die, and I think they're right. The hardcover book is a print artifact, a technique publishers discovered that would let them milk the same stone for blood not just once, but twice. They could sell a new book at a more expensive price point to early adopters and then release a cheaper version to the mainstream months later.

Digital books don't allow this. They help democratize content. On the plus side, this means readers can—and often expect to—pay low prices for premium content, without hemorrhaging money into a hardcover edition. But on the minus side, a revenue source for publishers is falling by the wayside, which means it's harder for them to take gambles on new content.

A publisher often has a portfolio of books, much like your own investment portfolio. Some titles are low-risk but low-sales, like bonds, while others are likely to be high-risk and high-return, like stocks. The ebook revolution may be making publishers think twice before taking on the risk of promoting a new author. But then again, perhaps this financial pressure will force publishers to take different risks in the digital space, to innovate new product experiences.

And this is healthy, even though some publishers will have to tighten their belts and others may fold by taking ill-timed risks that are too bold. It's healthy because the culture of reading itself is changing.

You used to be able to go to any college town, hang out in the college coffeehouse, and see people reading at their tables. But now it's different, which should be no surprise to you if you're a college student or the parent of one. And believe me, I know—every time I traveled on Amazon business to visit publishers, I would stop at a college town and check out the campus bookstore and coffeehouse.

I would come to the coffeehouse expecting revolution, expecting to see people reading *The Communist Manifesto* or at least sci-fi novels, but now I see people sitting at their laptops reading Facebook or watching YouTube. Though there's jazzy music on the coffeehouse radio, everyone is listening to his or her computer on earbuds. Books are at best decorations in the coffee-shop windows, encrustations of a former function that coffeehouses no longer serve.

In a broader sense, that's true of books in our culture. Books are becoming decorations. Whither the printed book? Indeed, it has withered. There are cobwebs in the corner of the coffeehouse and spooks like me who sit on the paisley couches, watching people alone at their tables on Facebook. Reading was never really a social experience, so I'm not alarmed by this. In fact, I'm happy, because I know that as digital books grow, more and more people will start reading books again in coffee shops—and I mean actual reading, not the indiscriminate snacking of bits and bites from the internet.

Of course, the downside of this is that people may start to expect digital books to behave like the internet, like a repository of content that you can snack on. Because it's true that, as you surf the internet, you're snacking all day instead of eating a full meal. A book is a full meal, and like any meal, you have to be willing to spend time preparing it and savoring it. True, it takes time to read a book, and it doesn't matter whether the book is physical or digital. The investment of time it takes to read and consume a book will remain constant regardless of the book's format.

I remember that in Woody Allen's movie *Sleeper* there was a machine called the Orgasmatron that people went into just to have orgasms. That's it—no sex required, no foreplay, no nothing. Until someone invents an Orgasmatron for books, where you get all the information you need in an instant, you'll still have to invest time in the experience of reading.

Moving beyond books, it's a small step from the written word to images themselves and a vast project to digitize all the art in the world's museums, whether it's boxed up in storage or hung up in plain sight on the walls. Humanity is smitten with the digital, and there are missionaries who will see to it—partly because of profit and partly because of evangelical fervor—that all these analog artifacts are digitized as hi-fi reproductions of the originals, and they'll be happy to sell them to you.

Is it too futuristic to imagine hanging an iPad in a gilt, rococo frame on your living room wall and seeing a high-quality selection from the Metropolitan Museum of Art displayed on it, pictures rotating every ten minutes? We can already display family pictures from our own digital photo albums, so why not display world-class art in our living rooms as well?

I think this is highly likely. In fact, the mass digitization of our culture could in some ways be considered part of a greater spiritual project. This wholesale conversion of the analog into the digital, of base "gold" into even more ethereal electrons, could be seen as part of a project that started centuries ago. It could be seen as part of humanity's dream of infusing all of matter with soul, a dream at once ancient and yet science-fictional.

After all, this is where the Web 3.0 movement itself is going. Your clothes will be computational devices. Your e-reader will talk to your smartphone and your scale and your coffee machine, and they'll all keep tabs on you, sense your mood, and recommend things for you to do or read or buy. Who would have thought that this dream of infusing the inanimate with the animate, of matter with soul, would ultimately benefit advertisers the most?

Bookmark: Altered Books

There were once so many passenger pigeons in America that their flocks darkened the skies for hours at a time as millions flew overhead. Ranging from the East Coast to the Rockies, they were a highly successful species of bird. Their fossils have been found as far back in time as the Pleistocene period, the same era that saw saber-toothed tigers ranging through what's now Los Angeles, or wooly mammoths roaming through Chicago, or giant ground sloths, ten feet tall, loping through what would become Las Vegas.

That we had giant sloths in America surprises me. That all the passenger pigeons died out stuns me simply because of the reason for their extinction. They were killed for their meat, and generations of Americans knew no other meat than pigeon meat. Tiny pigeon sausages. Pigeon pies. In the span of about a hundred years, one of the most common American birds was gradually exterminated.

Likewise, books once had a glorious range. Books roamed the world. They traveled in luggage on the Pan-Am flights of the 1950s, were carried in purses and satchels on trans-Atlantic schooners, were carried by the Pony Express across the continent, and were often an important part of dowries of noblewomen. But now the range of books has shrunk, like the range of passenger pigeons, although not yet as terminally.

You can still find books on dusty ornamental bookshelves in some hotel lobbies. You can find them in the lost-and-found bins of large train stations and on sun-bleached shelves at beach resorts. You can still find thriving populations of books at college bookstores and libraries. Somewhat surprisingly, you can also find books in prison libraries, which boast higher circulation rates than almost any other kind of library.

But like passenger pigeons, like coyotes, like black bears, and like ancient coelacanths—dinosaur fish from millions of years ago that only live off two islands in the Indian Ocean—books inhabit a restricted range when compared to how prolific they were in their former glory days.

In ecological terms, books are threatened with extinction.

Books aren't capable of reproducing in the wild or in captivity, although more and more book titles are published every year. But fewer physical books are being sold with every passing year, even if there are more titles to choose from. Book sales overall are tipping toward digital now.

Books are threatened with extinction, but like the smartest of animals in the wild, they're adapting. They're evolving instead into ebooks.

It's almost contradictory for me to be a futurist of books. It's like being a futurist of telegraphs or a futurist of rotary phones, because the death knell for print books has, in my opinion, sounded. As printed artifacts, books share a sacred reliquary along with eight-track tapes, gramophones, and LaserDiscs. But print books haven't died yet, and they're not going to go gently into the night.

In the upscale home décor stores of the future—and by future, I mean ten years from now—tucked in among the rugs and tapestries and oversized urns and stuffed animal heads, you'll start to find books sold as decorative items. They might be artistically bound with strips of copper. They might have keyholes from doors installed onto their spines. They might be artfully aged and lacquered. Perhaps they'll be set up on pedestals, or a small ceramic pigeon will be perched on the book. But you'll start to see books altered, turned into art objects.

In a trip today to my local art town, I saw three stores that sold these types of altered books. Some of the books were turned into pulp and molded into trees, with smaller books hanging from them as fruits would. I saw pages from a book carefully razor-bladed out with an X-Acto knife and painted to show scenes of children playing in a field.

What does it say about us as a culture that we're turning books into art?

To me, it says we're aware of the passing of books, and we're mournful. We feel pent-up nostalgia for books. We're aware of a genuine loss, one that we can only express with X-Acto knives and spray cans of lacquer and glitter. We're altering books, making them into art and ennobling them with ideas that are too hard to put into words. We're transforming humdrum leather-bound books that were formerly commodities into artistic statements.

We're aware somehow that art will last longer than commodities, and our artists are salvaging some books in a repurposed form, in the hope that some of them will last through the ages. Because let's be honest: do you really think the major libraries are going to hold on to all of their print books in an age of cheap terabyte hard drives? Do you really think the Library of Congress is going to digitize its book collection and then keep all the print books once they're digitized? They won't, and how can they? There's simply too much material to store.

So there's going to be a massive die-off of print books. They're not emigrating, flying overseas with the sound of pigeon flutter as their pages loft them through the sky. Books are dying. Future archaeologists will speak of the Gutenberg Era and the sharp discontinuity of our time, characterized by a major extinction event that has left no print books in the fossil records.

The artwork of Georgia O'Keefe often depicts bleached skulls on a desert landscape. When artists get around to painting the end of the Gutenberg Era, they'll perhaps paint the bleached books left behind on the literary landscape.

I think highly of printed books, but I already think of them as bone-white, bleached, inert, and dead—unlike ebooks, which seem to sparkle with electricity and wonder. I look at my walls of printed and bound books like they're all polished skulls in a curio cabinet. Just as I'm achingly sad to see them go, I'm also excited at moving onward into the future, into the digital.

But what about you? Have you come to terms with the death of printed books? Have you grieved, in your way? Care to share your thoughts or help others through the mourning process?

http://jasonmerkoski.com/eb/21.html

Reading: A Dying Art?

I can't appreciate fiddle music. No matter how good it is, fiddle music sounds to me like someone plucking the guts of a sick cat. But I know, rationally, that there must be truly great fiddle players. My mind understands this, even if it can't appreciate that kind of music.

Some things are simply matters of taste. Cilantro. Sushi. Cuban cigars. Krautrock. Spiders. There are no doubt items in this list that you find distasteful. And perhaps some that you appreciate, as a connoisseur might. Your taste for these is partly learned. In our country, we have developed an appreciation for sushi, for example, which is essentially raw fish. Spiders, by contrast, rarely make it onto the haute cuisine menus of our restaurants.

Culture is shifty, and as anyone who has traveled outside his or her home country knows, it can vary widely. And yet, some parts of culture are universal.

We all have an innate sense of storytelling, for example.

Whether you look at the oral culture of the Homeric Greeks, or the stories of the Navajo, or the stories of Jonathan Swift or Charles Dickens or any contemporary author, you'll find that most stories deal with people. This should come as no surprise. As people, we care about other people. It's part of our tribal ape heritage. It's wired into us. We're programmed by patterns in our own brains to care about people, to find them fascinating, and to see them even when they're not present, like ghost lights in the dark.

As an example, consider *pareidolia*. It sounds like a disease, but it's the surprisingly common tendency we have to see faces where none exist. And not just any faces—not bear faces or panda faces or fish faces—but the faces of people. We see them in whorls of wood and in the clouds overhead. There's even a shrine to a tortilla in southern New Mexico. If you look hard enough at the tortilla, you can see the face of Jesus. We have triggers that fire when we see things that resemble faces. These triggers sometimes misfire, hence pareidolia. Looking for faces is clearly important to us because it's biologically programmed.

Our sense of story is just as innate.

Good stories work well when they engage us in what we care about. Fiction does well when it paints a clear picture of a person, outfitting him with a camel-hair coat and a red beard. If the picture is too abstract, we don't engage. Likewise, a cookbook will fail to make us salivate if it doesn't have a photo of a pastry drizzled in chocolate sauce or a glistening sirloin steak cooked to perfection.

This preponderance of detail is what makes books work best. It takes a special kind of reader to enjoy Samuel Beckett and his abstract, disembodied fictions. We need details. Details resonate with us. Or, more properly, they resonate with our imaginations.

As far as I know, no clinician has isolated the imaginative faculty. It can't be seen in any anatomy book. There are no brain labs at Harvard where rabbits are being vivisected to find the elusive imaginative faculty. It can't be removed with forceps or pinned to a Styrofoam dissection tray. There are no crackpot scientists posting papers about the imaginative faculty in the pages of *Nature* or online sites like arXiv. The imaginative faculty cannot be bottled like a freakish two-headed snake in a bottle of ether at a carny sideshow. In fact, the imaginative faculty resists my own attempts to describe it, which is why I can only say what it isn't.

But we all have this imaginative faculty. One theory for why we have this relies on evolutionary psychology, which offers explanations based on how our original human ancestors might have experienced life in the savannahs of Africa. If you believe in evolution, this theory might explain the imaginative faculty as an extension of being aware of predators and prey.

In the savannah, you would have to be alert to lions or tigers. You might imagine a tiger approaching if you heard a twig snap in the dark, and you'd react accordingly. Likewise, as a hunter, you would have to put yourself inside the head of your prey to capture it. You would have to think like a wildebeest, for example. Put yourself into her hooves. Anticipate how she might react, which rocks she might jump over, which trees she might try to hide behind.

In this sense, hunting requires storytelling. And because imagination is linked to our very survival—the ability to eat and the fear of being eaten—this faculty may have developed over time as an evolutionary adaptation, linked of course to our large brains.

However this faculty adapted, we can use it to put ourselves inside a really good book. Your second-grade teacher might have helped you to develop an appreciation for books, but it was always innate inside you. I would even bet you first discovered your imaginative faculty while reading a book when you were a kid. Perhaps it was a fantasy book about wizards of angelfire who fought dragons, or a comic book about Krypton or Eternia, or a story you imagined of Bible heroes flying through the sky.

The imaginative faculty is part of our human condition. We look for patterns and apply them to ourselves. We read a book and patch the details provided by the author together with those from the circumstances of our own lives. Whose face does the man in the red beard wear? Your mind fills the gaps with details from your own imagination. Perhaps you see the face of an old professor behind the red beard. An author doesn't need to spell out all the details when he writes. He can rely on you, as a reader, to fill them in.

You patch these details in with your imaginative faculty, just as you sometimes see faces in knotted wood.

This faculty is innate, but it can be improved by training.

One doesn't go straight from reading *Dick and Jane* books to Beckett's *Waiting for Godot*. But we do become gradually more voracious readers as our critical thinking skills improve and as we learn to look for nuance and ambiguity. We learn to crave details that get gradually more complex and characters that are less black and white. As for words, we come to crave the occasional neologism.

We crave, in fact, the fullness that experience itself can bring. And when we can't get it from an author's own words, we patch in our own experience. When you read a book of fiction, you use details from your own life to fill in the author's missing gaps. You caulk the author's stories with scraps from your own outlook and knowledge.

Reading demands a lot out of you. Out of all readers, in fact.

Sadly, reading rates are dropping, although existing readers are not giving up the reading habit. Once you're a reader, you're always a reader. What's happening instead is that fewer people are developing the reading habit every year. It takes time for a child to develop this imaginative faculty to a point where it becomes rewarding. It takes time for the feedback loop to kick in. There are not enough new readers buying books every year, which is a matter of population dynamics. For a population to grow, there have to be more net births than deaths. Unless this decline is arrested, reading will decline.

I could stand on my soapbox on ten thousand street corners, talking about how important reading is, but it wouldn't help. I could have my own TV show that teaches kids about reading, with LeVar Burton and Justin Bieber and a masked Mexican wrestler, and it still wouldn't be enough.

I could airlift a million copies of *Dick and Jane* over the poorest part of Appalachia, the area with the country's lowest literacy rates, but that wouldn't help either. Against the onslaught of digital media, reading may decline to nothing more than a faded art form, neglected like ballroom dancing or Appalachian fiddle music.

In this sense, you might think that the future of reading is doomed. How can reading cope, given that movies and TV shows already provide a surfeit of details for us to work with? When you watch *Star Wars,* you don't need to imagine what Darth Vader looks like under his mask; you can see each lurid scab on screen.

Likewise, video games don't make the same demands on you as reading. Animators have crafted a whole world for you, along with computer-generated faces and professionally recorded voices. This makes it easier for you to experience the movie or TV show or video game, as your mind isn't being taxed. But this is itself a drawback. If the imaginative faculty is seen as a kind of muscle that you flex inside your mind, then not using it may cause it to weaken and atrophy.

In some ways, this is a problem of philosophy. Does imagination matter?

If pre-imagined media experiences are what matter to you, then ebooks alone cannot compete against the onslaught of TV, movies, and video games.

Many ebooks are still mostly text, and the few experiments that attempt to hybridize movies and reading come off like tigers mated to killer whales. They're like bestial monstrosities. Interesting as such experiments may be, the future of books does not lie in this direction.

No, the future for books is a return to the imaginative faculty, to the resonance between reader and author that causes the reader's heart to flutter and his pulse to quicken, which causes him to sympathetically sweat when a zombie crashes onto the page or when a loved character is brutally murdered with a knife through the eye. Movies and TV and video games may win out in terms of production costs and special effects when compared to a humble book, but no movie yet made can let you into its world. Readers inhabit a book. They burrow into Frodo's hobbit hole and curl up with him for a pot of tea. In contrast, the only way to "read" a video game or movie is when you are not participating in it.

As an example, I'm on an airplane now, heading back to Seattle. As I walk down the aisle to stretch my legs, I see plenty of Kindles. It sometimes seems like there are more Kindles on airplanes than Rollaboards. But even with all the Kindles and iPads, books seem to be outnumbered. On this airplane, at least, there are more laptops and video-game consoles, more people playing games and watching movies. The written word is outnumbered two to one.

When I return to my seat, the kid next to me is playing his video game. He's utterly absorbed by the blinking dots, hunched over his game like Quasimodo and reacting to the electrons on his screen. It's reactive. It's a matter of stimulus and response. And I know this feeling well; I'm no stranger to video games. I know that when you're absorbed in a game, it's all-consuming.

But afterward, when the game is turned off, you can reflect, strategize your next steps, and plan ahead. It's at such times that you really "read" a game. And likewise, the most voracious "readers" of a movie are the fans that obsess about it afterward, who imagine themselves as characters in the movie, or who buy books or director's cut DVDs afterward to read into the nuances of the movie's world.

I think this redefinition of "reading" bodes well for the future of books. But it means a shift in thinking. It means that any media experience can be "read" like a book, that there's no preferential treatment of books over other forms of media, as long as the content is "read" with an active imagination. Because philosophically, I do think the imaginative faculty is important. I couldn't live without it.

And I think that most of the successful people I know at Amazon, Apple, and Google, as well as among the publishers of the world, are those who are most creative, most imaginative. These are people who "read" into experiences, who don't just talk to me about what was on TV last night but who imaginatively transplant themselves into the worlds of those TV shows. They're the kind of people who wonder what it's like to be a Cylon in *Battlestar Galactica*, who "read" into a media experience and apply it to their own lives, and who patch in details of the media with their own life experiences to personalize it.

I think any piece of media is capable of such a reading. A movie doesn't have to be a classic like *Citizen Kane*. It could be anything, as long as you resonate with it and read into it with your imaginative faculty. Because books demand this form of reading, they're here to stay for a long time in print or digital form—at least for that select group of people who enjoy imaginative reading.

For such people, books demand to be read. And they demand your attention. And paradoxically, because books by their nature aren't as visually or auditorily rich as other forms of media, they engage our imaginations more strongly because of our need to patch in details. It's a wonderful feedback loop: the more we read, the more we need to read and the less satisfied we are by entertainment that panders to our senses but deprives our imaginations. Once a reader is hooked, there's no way to give up the habit.

As readers who are accustomed to the deep resonance we have come to enjoy when our imaginations are engaged, we're hooked. And ultimately, because the imagination is so innate, no simple technological silver bullet can be applied to books or ebooks to make people read more. It doesn't work that way. Reading—whether you're reading a book or "reading" a movie—is a personal act of volition, of attention, of mindfulness. Reading comes from within. It takes energy. But it's also so very rewarding. Reading is a gift that keeps giving.

At least, as long as you're able to pay attention to what you're reading.

Bookmark: Attention Spans

As retailers move toward tablets to let you consume all kinds of media, we're finding that our focus often gets diluted and our attention spans get—what? What was I saying? Hold on, let me check my email and do a quick tweet.

Don't get me wrong. I've paid more than my fair share of dollars toward Apple's billions in app sales. And I've done this with the drawback that when I do read on my iPad, I often find myself bouncing from the ebook application to the browser or to Facebook or a bunch of other applications. And the single-threaded reading experience that I get with dedicated e-readers or even print books is lost.

The reading I do on my iPad is more like snacking than eating a full meal. That wonderful faculty I have in my brain as I read, the way that my temporal and parietal lobes light up as I explore what-ifs and puzzle out the layers of meaning in the book I'm reading—well, on a tablet, those lights grow dim, and I lose focus. And I'm a fairly disciplined guy, so it's not a matter of my own susceptibilities. These multifunction devices, which will be a core part of our future, engender a less focused mode of reading.

This is problematic, because as your mind wanders like a moth at a carnival after sunset, bumble-flitting from booth to booth, from light to seedy neon light, it may never return to where it began. Now, this flittering has the side benefit that if we can channel ourselves properly while reading, we'll be able to use other applications as adjuncts to look up words or to go online and find out the hidden meanings or subtexts. But ideally all this functionality would be seamlessly present in the reading experience itself, so we wouldn't run the risk of losing our place in the reading or our train of thought.

So we either need better applications that keep us rooted to what we're reading, or we need to police ourselves—perhaps with lockouts that we apply to ourselves that prohibit us from wandering out of the book to check our email or surf the web or only allow us to do this once an hour while reading. Perhaps future software updates for the iPad will allow teachers to lock devices down into ebook-only mode or give students intermittent access to the non-ebook parts of the device. Lockdown controls like this would probably be useful for a lot of adults I know too.

Lockdown controls aren't the only thing we could benefit from. I think we can all benefit from a brush-up course on digital hygiene, on learning how to focus. And I think we're going to learn just how important social networks are. A 2012 study by the Association of Magazine Media showed that Gen Y was reading more magazines than ever, although this reading was tied to an increase in the use of social networking sites. So let's face it, ebooks are going social, and it's going to be a strange symbiosis, like that between a hummingbird and an orchid: one without the other would likely not last. Digital books will form an unlikely alliance with social networks, and they'll both survive the changing tides of fashion and the flighty whims of technology.

Still, one of the reasons I adore dedicated e-readers like the Kindle and the Nook, as opposed to tablets like the iPad, is that they keep your attention on an ebook as you read. Like with a print book, you've got a dedicated reading experience with no distractions—no buzzing lights or videos or ads for meeting singles online or tweets to respond to. I worry when reading experiences start to include too much distraction and context shifting. As someone sensitive to media ecology, that's where I draw the line. I think all of us, our children included, should be encouraged toward dedicated experiences, not distracting ones.

Our devices are shortening our children's attention spans. Our children need to concentrate when they learn to read to become good readers—and from that, good thinkers. But our hypermediated environment is one of constant distraction, so our kids are often learning to read—and through that, to think—in a rather shallow and careless way. It never used to be possible, let alone culturally acceptable, to read and watch TV at the same time. You would have to pick one or the other to focus on.

But now with devices like the iPad, you can multitask between them, switching from reading to watching a video when the book becomes too hard. And let's face it: our brains are lazy. Ask any cognitive neuroscientist, and they'll tell you that our brains are machines for avoiding work, if there's any work to be done. And reading is hard work. It's rewarding, true, but you have to actively work at it. When you skim a book and passively read it, you don't recall as much of what you've read as when you pause, linger over sentences, find the humor or irony in them, and actively work at the reading experience.

That said, not everyone can focus. Attention-deficit hyperactivity disorder (ADHD) causes inattention, distractibility, and disorganization. Incidences of this disorder are on the rise. In fact, it's estimated that up to 10 percent of American children have this. At the time of this book's writing, doctors still don't know what causes ADHD. But most doctors would agree that you don't try to fight inattention with more inattention. If anything, children with ADHD are encouraged to create routines and avoid distractions. Snacking on digital media on iPads and similar multifunction tablets only feeds the inattention.

Not just children have ADHD. Many adults do too, and the numbers are still climbing. Maybe it's part and parcel of carrying around so many smartphones and tablets and laptops, of being too plugged in to the internet and chat windows and glittering digital eye-candy. But this disorder is debilitating. In the end state, if this continues unchecked, we run the risk of becoming a nation of ADHDers, unable to focus, engage, or reason clearly.

What's the way out of this?

Simplicity, mindfulness, and attention. It might be as simple as doing nothing. As long as it's the right kind of nothing.

In the book *Hamlet's BlackBerry*, author William Powers describes a technique that works for him. He calls it a "Walden Zone." It's a room without electronics. A room in your house where you can think, like Thoreau on Walden Pond. A place where you can meditate and contemplate—and ideally, you're not contemplating what your next game of *Angry Birds* will be like or how you'll beat your former score.

It's a technique I use in my own life. There's always a room in my house with no gadgetry, and I try every year to take a vacation for a few weeks somewhere without electricity. I try to reconnect with myself. Even if you don't suffer from ADHD, this might work for you too.

If you have other techniques to stay focused that work, why not share them with others who are passionate about ebooks but wary of the perils of having too many distractions? And if you're a parent or a teacher, what do you think about how reading is taught these days? Do you think kids can become good readers when music and TV and the web and texting are taking up their attention and taking them out of their books?

http://jasonmerkoski.com/eb/22.html

The Last Digital Frontier

One of the amazing things about the ebook revolution is how much attention it has gotten in our culture. Ten years ago, hardly anyone talked about the book-publishing business. Even editors and publishers were bored with it. Movies and TV were much more exciting. But today, you can find stories about the ebook revolution online and in newspapers almost every day. Why is this so fascinating to people?

I believe it's because books had a solidity to them. They represented the accumulated weight of our culture. Books were the last bastion of the analog. Prior to the Kindle, all other forms of media had been digitized. Music, movies, TV shows, video games, even newspapers were available on the web for instant download and instant gratification. But books remained in print.

But now, the last bastion of the analog, the last stalwart bulwark, has finally been cracked, and books are available for digital download. With the advent of ebooks, books will never quite be the same. Now, our eyes will grow accustomed to LCD screens and eInk displays instead of paper softly lit by glowing fireplace embers at night. Our kids will never know the subtle way that books get scarier by night as you curl up under the covers with a flashlight to read.

These next few years will be momentous for the book industry, as it shifts from a purely physical mode to a digital mode. But this shift into the digital is happening everywhere, not just with books. Everything we take for granted in the physical world is up for grabs in the digital world, including core concepts like ownership of ourselves and our creations, digital or otherwise.

As we transition our lives wholesale into Facebook and Twitter and communicate more with email than face to face, what does it really mean to own something or even to "be" in the purely existential sense of Hamlet in his soliloquy? What does it mean as our books shift to the cloud from our trusty wooden bookshelves and from neat or perhaps messy stacks next to our beds? What does it mean as our media—our books and songs and movies—are no longer real-world things with any substance that we can feel with our fingers? What does it mean as we move our memories online into social networking services or as we post our photos onto websites like Flickr instead of printing them out at the pharmacy and putting them into photo albums?

These questions persist and will only grow harder to answer over time.

All the papers, all the records and receipts of our lives, will go digital next. There'll be ways of browsing them, handheld devices that we can use to browse our own lives through these collections of bus tickets and love letters that once meant so much to us.

I think it's a stretch to say that we'll live out our lives entirely in the digital world like cyberpunk authors of the 1980s would have had you believe, that we'll sell off our furniture and live instead with bare-bones lamps and beds made of origami that can be crushed underfoot when they're no longer needed, that we'll live in small shacks like U-Stor-It lockers, jacked into computers, and that we will only care about our avatars and the clothes they wear. It's a stretch to think that we'll live this way, but nobody knows. What it means to be alive in a digital sense is still up for grabs.

Recently, I've been reading a lot about the seven wonders of the ancient world. In particular, the fact that some still exist. I'd thought that the Pyramids of Egypt were the only wonders of the ancient world that still remained, but actually, there are remains of almost all the other monuments from antiquity. There are chunks of masonry of the Lighthouse at Alexandria in the Mediterranean Sea, remaining from when the lighthouse was toppled by an earthquake. There are fragments and sculptures from the Temple of Artemis held at the British Museum after being recovered by early archaeologists.

There are still ruins of the basement levels of the Mausoleum of Halicarnassus, and you can go there and walk among the ruins of an ancient wonder. It's even said that the base that supported the Colossus of Rhodes still survives at a church a mile away from the bay where the Colossus was said to have once towered. And who knows? One day, there may be a cuneiform tablet unearthed that contains a plan for the Hanging Gardens of Babylon.

Though the Old World has been scoured by archaeologists, they're always turning up new things. The Statue of Zeus at Olympia, the last of the seven ancient wonders, no longer survives, but the workshop where it was built was recently uncovered.

I'm surprised that remnants of the ancient wonders have persisted for millennia, and I'm encouraged, because if brute stone can survive, surely a digitized person can survive, as well. We should all be able to float down the eons in a Pharaonic funeral boat, immortal in a way that the ancient Egyptians could never dream of.

Perhaps I'm in an elegiac mood these days, but I wonder what would prompt people to build a digital version of themselves. Could I build a digital monument to myself, a digital Mausoleum of Halicarnassus? Would it be something that merely baffles my friends and family, or would it live on as a testament to a madman's desire for immortality, as mad as the Pyramids or a forty-foot-tall silver statue of Zeus?

Could a version of my digital self become a companion for people in the future, a confidante, someone they could talk to? Could my digital self gradually learn from itself or others and subtly reprogram itself in the same ways that I myself might? Will there be a place where digital personas can congregate together, some digital mortuary grounds or Second Life where they can talk about who they once were or argue about ideas?

I'm not sure. But I've got the mad Pyramid-architect desire to try and find out, to see what happens, one way or the other. A digital self, if it can avoid bit rot, is a kind of immortality. It's the oldest dream of them all, the Faustian dream of living long enough to know and observe everything. Except that there's no devil in this Faustian pact—or at least, no devil that I'm aware of yet.

In the telling of the story, the devil granted Faust all the knowledge he wanted, with the catch that his soul would one day be claimed by the devil. The knowledge was limited only by death, which of course explains the motivation for Faust to cheat the devil and live on. Sadly, in both Christopher Marlowe's and Johann Goethe's versions of the story, Faust inevitably dies.

His tale is ultimately a moral one, the message being that we can't live forever. We can't come to know everything. And that's fine. I know I won't live forever. My body and mind are frail, just like yours, like everyone's. But that's okay. Because surely through my digital self, I'll live on—right?

And this, in fact, is the final digital frontier. The digitization of memories and minds themselves. It might take a hundred years before the heirs of Jeff Bezos and Steve Jobs figure out how to digitize human brains and make them available for purchase and download. But once that happens, it would be an amazing experience to download the personality of your deceased grandmother and to speak to her for a few hours. Or perhaps you could have a conversation with yourself as you once were. Or speak with any of the great minds of history and have a dialogue with them or argue with them.

For example, I'm reading a classic sci-fi book now called *Martian Time-Slip* by Philip K. Dick, and I'm stunned at how good it is. If I could download the author's personality and start talking to him about his book, I'd feel overjoyed. The closest I can come now is to talk to other readers or post to the dead author's Facebook fan page, but it's clearly not the same as a genuine conversation with the author.

Of course, in all likelihood, the minds of wealthy entertainers or technology early adopters will be digitized first. Theirs will be the minds available a hundred years from now as public domain recordings for people to download for free. And while theirs will be the first minds to be digitized, the quality will be poor, like that of wax cylinders or early ebooks.

Though created with the best that technology could once offer, they'll eventually be seen as grainy, more lo-fi than other hi-fi brains available for download later in the future, so they'll be relegated to public domain archives that hardly anyone ever visits, the equivalent of the Department of Special Collections at the University of California.

And who knows, maybe Jeff Bezos will convert one of his data centers into a building to house his digital brain. Heck, if I had the money, I would do this too.

You don't have to take my word for this; you can read any contemporary sci-fi book to see the same insights and impulses toward living digitally in a disembodied way. Because these ideas are now part of our culture's currency. But for now, you and I are as analog as it gets. We get hangnails and wrinkles on our feet. We drink entirely too much beer and suffer entirely too much of a hangover the next day. It's all part of being analog, and there's nothing wrong with it. In fact, I like it—wisdom lines, headaches, and all.

Perhaps I only like it because I have no choice in the matter, and I choose to look the other way when I suffer stubbed toes or pimples in ungainly places like the insides of my ears. Or perhaps I like living in the analog mode because of all the great feelings I can experience, what cognitive scientists call "felt states"—the sun on my skin, the taste of a fresh blueberry, or the wonderful, fresh smell of a spring morning. I'm as happily analog as I can get, and I will be for a long while.

And though I may be embodied in an analog form, I can still read great digital ebooks.

Those who read this in the future may sometimes forget that books weren't always digital. They may look back upon us with disdain because we don't have brain implants to post live Twitter updates. They may look down upon us because we have sex with one another instead of using electronic Orgasmatrons. They may frown upon us with the face of history because we're no more than apes who type software and emails with fingers of skin and bone, because we're pitiful creatures who wrap rags around our frail bodies as we walk to and from work.

I can only plead with those in the future who read this to remember that if not for us, there would be no digital books today, and the future would be less rich and nuanced. If not for us, future readers wouldn't be floating as brains in an etheric vat, surrounded by digital books and videos and music as they sample from all of human culture like it's one vast buffet for the mind. Those who read this years from now, please don't forget that the future wasn't always digital and that books weren't always electronic.

Because without the ebook revolution, the future could never have happened.

The ebook revolution is the story of a small group of people who set out to change the way the world reads. I mentioned before how I found it at once eerie and amazing to be in a meeting with Jeff Bezos, scrutinizing the number of lines that should appear on an ebook's page, because it was the same kind of thinking Gutenberg used more than five hundred years earlier. And like Gutenberg's team, this small group of people at Amazon worked their way up from square one to reinvent reading. And we succeeded. Reading has not only been transformed but also rebooted.

But this success came with a cost. There were unintended consequences of this success, which meant that many ebook features had to be shelved at Amazon. It became important for Amazon and other device makers to keep up with their competition, which meant that certain innovative features were deprioritized so that resources could be spent on the arms race of keeping up with competitors. These ebook features will eventually be built; I'm not worried about that.

Amazon launched the ebook revolution, but now, the future of books is being tended to by people outside Amazon's walled garden. By innovative publishers or venture-capital-funded startups or iconoclastic propeller-heads. Innovation is out in the world now. It's out of the hands of Amazon and other technology giants. I believe the smaller, more nimble, more purpose-driven groups will succeed in building these features out. And of course, as always, the readers ultimately win.

Innovation sparked the ebook revolution. And while companies like Amazon and Apple are now raging bulls whose horns are locked in competitive combat, that innovation has gone out into the world. Publishers, as well as authors, are able to innovate. Readers themselves can innovate.

We ourselves, as readers, can reshape the future of the book!

We can reengage the way that books work in our own lives and fan the flame of reading again. In my generation, books have lost people's attention spans, lost them to TV and movies and video games and the internet. But now books are being revitalized. Reading has never been more interesting, and it's all thanks to ebooks.

I said earlier that, to the brain, there's no difference between the words in a book and the words in an ebook, but ebooks introduce us to more than just words. They introduce us to other people and let us talk to our friends and family, right in the margins of an ebook. New life is being breathed into reading by the ebook revolution. If reading can be saved in our ADHD culture, then it's thanks to innovative ideas from Reading 2.0 and to pioneering publishers and retailers and tech startups large and small.

I'm happy that I had a hand in making the ebook revolution happen. I did a lot for digital books. I turned the Kindle flywheel a few turns. I turned a page of the history book, turned books into ebooks. I burned the printed page and fanned the flames, helped to kindle a revolution.

I'm happy to have participated in the Kindle, along with a bunch of others. They're people I sometimes miss and sometimes don't, but all had charisma and character. There were enough characters on the Kindle team for a new font. At the very least, there was enough of a cast on Kindle to make a movie about it. BlackBerry-addled vice presidents, Jeff Bezos's endless array of executive assistants, and engineers with barbecue stains on their shirts debating death matches between killer whales and tigers. They were all part of this ebook revolution, all part of something—well, something magical. Something revolutionary. As revolutionary as the invention of books themselves in Gutenberg's day.

We don't know what it was like in Gutenberg's workshop, of course. But we can imagine the darkened interior, perhaps see the lampblack and soot. We can imagine the sounds of molten metal being poured into molds or the occasional scream as someone scalds himself, a spattle of molten metal on his skin, or perhaps it was someone's finger that got caught in the printing press. We had it easier at Amazon; there were no molten metal mishaps, but we kindled a revolution nonetheless.

What we achieved will have consequences in the decades and centuries to come. But inevitably, Amazon will become just another name in the pages of the history books—or rather, history ebooks. Perhaps in the future nobody speaks of Jeff Bezos or Steve Jobs. Perhaps in the future, corporate entities are treated as people and our history is written by the likes of Apple and the Internet Archive and Google. Perhaps history is written about corporations by corporations. And it may well be in such a distant future that nobody really recalls that people like you and me brought digital books to life, that we made this quiet, bloodless revolution happen. That's fine. We did it, even if nobody else will know about it.

We did it. And I wish that for just once in corporate culture, people took time to celebrate in a human way. I wish that just once Jeff Bezos brought us all together in a big ballroom, everyone on the Kindle team. I wish that just once he said nothing about flywheels, nothing corporate.

I wish that just once we all paused to celebrate what we achieved, that we wordlessly reached out to one another, held each other's hands, and laughed like children, dancing in a circle. We could put aside our paychecks and all the politics, put aside our differences, and everyone could simply hold hands and pause before bowing to the audience, as this chapter in the history book closes, like a curtain falling over the stage.

And in this dream I have, it wouldn't just be Amazon people. No, in the ballroom next door there'd another party where everyone from Apple has come together. Everyone who cared an ounce about ebooks and the iPad are likewise celebrating. No more corporate platitudes or PowerPoints, just corks popping, wine flowing, people eating and dancing and laughing and just plain celebrating with all the honesty of early-childhood innocence. And yes, the next ballroom over has employees from Google, and the next one has everyone who worked on the Nook. Everyone is celebrating, everyone past and present.

We all stumbled onto a great thing with ebooks. But we're all echoes, opportunistic echoes from an earlier time, all the way back to Gutenberg's time when books as we know them were first invented. In my mind, as visions of ballrooms fade, I can see, out on the landscape beyond, a summer's afternoon in the 1450s and a ghostly scene of Gutenberg and his workers coming out of his workshop. They come out with ink stains on their hands, smudged fonts on their shirts.

Maybe Gutenberg himself has a glass of blackberry wine or a stein of beer, raised now to celebrate the first Bible that started it all, everyone celebrating these thick-inked books that now soar into the clouds.

http://jasonmerkoski.com/eb/23.html

Acknowledgments

If I could enshrine anyone on my Facebook wall and follow them forever, it would be these people.

I want to thank Dominique Raccah, my publisher, for the vision and tenacity to see this book through. Authors rarely get to sit in a hip café with their publisher, but that's how this book started. I want to thank my editor, Stephanie Bowen, who transformed this book into a thoughtful gaze through time's periscope, forward and backward into the history of ebooks. I owe everything to my publicity manager, Heather Moore, for releasing this book into the world so successfully. I want to thank Jeff Bezos, Steve Kessel, Felix Anthony, and Bob Goodwin at Amazon, who taught me all I could ever want to know about innovation. I have to thank my father, Paul, who introduced me to the smell of newsprint and the word-sparks of great writing, and my mother, Kay, for encouraging me to read and play through all the summers of my youth. And finally, there's Hilary, who shared her books with me when we were hesitant teenagers; in the years since, we've shared so much more together. Relationship doesn't do justice to what we have; it's an elated relationship, so call it an elationship. And it's an elationship that sparks and burns brighter every day. Thank you, all.

About the Author

Jason Merkoski was a development manager, product manager, and the first technology evangelist at Amazon. He helped to invent technology used in today's ebooks and was a member of the launch team for each of the first three Kindle devices. Trained in theoretical math at MIT, he worked for almost two decades in telecommunications and e-commerce with America's biggest online retailers, and he's worked with publishers large and small to shape the future of ebooks. As a digital pioneer, he wrote and published the first online ebook in the 1990s. As a futurist, he's equally at home in Seattle or Silicon Valley, although he's drawn to the high desert of New Mexico, where he can string up his hammock and stare into the clouds and see ancient petroglyphs.